Promises of New Biotechnologies

Edith Breburda D.V.M., Ph.D.

Promises of New Biotechnologies

Edith Breburda

First Edition 2011

Scivias ™

PO BOX 45931

Madison, WI 53744

ISBN-13: 978-0615548289
ISBN-10: 0615548288

Acknowledgement

My sincerest thanks and deepest gratitude belong to Bishop Emeritus William H. Bullock, Diocese of Madison, WI, for his keen interest in this book, his encouragement and constant inspiration. Furthermore, I express my heartfelt appreciation to my friend Dr. Ruth E. Hurley, D.V.M., Ph.D. for her scientific feedback.

Contents

FOREWORD
William E. May, Ph.D.

Dr. Breburda's *Promises of New Biotechnologies* is a fascinating and important book on major issues in contemporary "bioethics." I put "bioethics" in quotation marks because Dr. Brebruda's excellent study does not examine and criticize the works on this topic by contemporary "bioethicists." The latter includes philosophers, theologians, lawyers and other non-scientists. Of these the most influential ones assume that membership in the human species has no moral significance, that "nature" is a human construct, and that the "new biotechnologies" and their creators can do a better job in "creating" plants, animals, men, animal-human mammals, and other entities than did the "Big Bang" eons of years ago. There is no need, in their minds, to appeal to any "god" or more than human source of meaning and value to explain "reality."

Breburda's book differs because the literature that she examines and on which she is an authority is by scientists at the forefront of the "new biotechnologies." A scientist herself, Breburda offers profound insights into and weighty "caveats" regarding the utopias "promised" by the developing biotechnologies.

In the preface to her book Klaus Küng wrote: "We are recognizing hybrids, stem cell research, clones and other technologies. Many admire the new techniques. Only few question them. Moreover, the world is proud about the outcome of the modern biotechnologies. They are appreciated as a sign of progress..." He then continued: "Nevertheless, the relevant question is what humans are allowed to do. The answer depends on the question what is a human being. The *only answer* is pointing to *our relationship with God* (emphasis added), our dependency on Him as His creation." These words in some ways express the main message of this remarkable book. I believe Breburda's "thesis" can be put this way.

> The "natures" of plants, animals, men, and all living things are not man-made constructs. They are the dynamic principles of these beings, given to them and making them to be the *kind* of beings they are by a loving, provident God. This God made man, alone of all material created things, to be the *kind* of being he is—a human being with a human nature—for a special reason. It is this. God wanted to create a being whose *nature* is inwardly open to the divine nature. Man's nature is the kind of nature that God himself could, should He choose, share Himself. And God chose to do so in the Person of his only begotten Son made man— the One who became "flesh" in the womb of the Virgin and suffered death to overcome the evil that man had brought on himself, and then rose to a new kind of life. And this God-Man Jesus Christ is the One who enables all men, united with him in baptism, to share his *divine* nature by dying to the "old" man and rising to a new kind of life, one that begins here and now.

This in my judgment is Breburda's counter-thesis to that of those who are in rapture over the "new biotechnologies" of genetic manipulation.

In the chapters of her book Breburda takes us on a guided tour of the "achievements" and "promises" of these biotechnologies. In doing so she discloses that an unintentional effect of overzealous British efforts to produce genetically enriched food for cattle was the Mad Cow disease. She shows how the artificial insemination of horses in the 18th century led to the development of the new reproductive technologies of *in vitro fertilization,* its permutations and combinations, the cloning of animals and efforts to clone humans, and how all this has resulted in the dehumanizing of women—valued now only because they produce oocytes—to various maladies suffered by the children thus manufactured, and other tragic consequences. She shows how genetically manipulated antibiotics have led to resistance to such antibodies with the consequence that new kinds of gastrointestinal, respiratory, and hematological diseases have sprung up, along with miscarriages and the increase of genetic abnormalities in mammals and men. I could draw this list out much further, but from what has already been said we can see what Brebruda does in the different chapters of her work.

Breburda's fascinating study is intended to help ordinary people understand the complexity and perils of the new biotechnologies and to remind us all of our responsibilities for the well-being of our human descendants and the environment in which we live. One hopes that it will have a wide audience.

Preface

Bishop Dr. Dr. Klaus Küng, MD, Bishop of the Diocese of St. Pölten/Austria

We are recognizing hybrids, stem cell research, clones and other technologies. Many admire the new techniques. Only few question them.

Moreover, the world is proud about the outcome of the modern biotechnologies. They are appreciated as a sign of progress, as proof what human beings are capable of. Nevertheless, the relevant question is "What humans are allowed to do?" The answer depends on the question "What is a human being?"

The only answer is pointing to the relationship of the person with God, our dependency on Him, as his creation. Humans are limited, but also responsible for their actions. Based on that background Edith Breburda is answering the questions we have regarding modern biotechnologies. Because of her profound knowledge in this area, in a humorous, exciting and easy to understand way, she provides true assistance to the reader bombarded by a variety of current opinions. May this book find wide acceptance which permits the clarification of the most relevant questions and problems in our time.

1 Phytomera

Imagine that your supervisor is standing in front of you. Excitedly he begins to tell you about Phytomeras. With each word he says, you become more and more confused. To be honest you have no idea what he is talking about. Right now, it is still a monologue, but what if he wants to know your opinion. What would you say? Is he kidding, you are starting to ask yourself. It does not seem very intelligent to confess that you have never heard the word, Phytomera. You get that blank expression on your face as if you spot a tornado. Your blood freezes in your veins. Then he stops talking and looks at you full of expectations. Does he sense that you feel uneasy and totally uncomfortable? In an instant, you are thinking of your response. However, you are so scared you are not getting any useful thoughts into your brain. You look at him; try to smile and then you take a deep breath. In that split second, as you try to get a word loosened from your extremely dry throat, his cell phone rings. What an unbelievable relief! It is your chance to leave the scenario. In the next hours and days, you try to avoid the conversation or even the contact with your boss. Still the word Phytomera has poisoned your guts. Constantly it is on your mind and you wonder what it is.

Phytomera sounds like a utopian word. Or might it be a term we should know from science fiction novels? Does it concern lives on another planet? Or does it inform us about an outstanding breakthrough in research? Is it a "rediscovery" of an ancient healing method? It could be an herb with incredible benefits and attributes. The effort of researchers finally paid off. Again, we gained information about an up until now unexplored ancient Chinese or Peruvian health secret. An universal plant ingredient used as remedy in the treatment of Cancer, Alzheimer's, Parkinson's, and Diabetes. Thanks to scientists, Phytomera is made accessible for the normal mortal. The substance and its effects were possibly known in the Middle Ages to naturopath-well-informed St. Hildegard, Philippus Aureolus Paracelsus, or Pastor Kneipp.

Unfortunately, a misguided, ignored and/or wrongly used dose removed the plant substrate from the healing list. Phyto is the Greek name for a plant. Might "mera" be associated with Chimera? Do we have a connection with this word? Does it remind us of our parents, or grandparents and their endured Greek lessons during their education? Do we combine with it a mostly forgotten language, or Greek fairy tales? Until today, Greek or Latin is the language we might gladly combine with masterpieces of music, as well as with the church language. Is there a connection between a language and religion, like German for the Lutherans? However, who in our enlightened times, in which religion is better not mentioned at all, is interested in that? It is a "foot in the mouth term" to speak about religion in modern society. Still Chimera is a superstitious word. Maybe it is describing some alien creatures. Whereas, we really believe in their existence. Besides its archaic sound this word seems somehow familiar. Do we associate it with immortality, eternal life, a cure for every disease and the elimination of all suffering? Do we remember those words spoken in modern biomedical research? They may resemble religious terminology even though they are used separately and have nothing to do with our religious belief. Chimera describes a half-human and half-animal composition. For example, the centaurs (from Ancient Greek) are a race of creatures composed of part human and part horse. In our times, Chimeras are associated with cell-based human-animal individuals created for modern scientific research. Phytomera must be a natural combining a plant and a mammal.

This is something no one ever spoke about except John Ronald Reuel Tolkien in "The Lord of the Rings" trilogy. However, that story is catalogued as pure phantasmal literature (Tolkien 1954/55).

Slang used by teenagers is a source for words like cabbage-head. In this context, the youth jargon is full of nonsense vocabulary. Other languages use words like "lettuce-head" or the beat will be cut off, which means you will lose your head, or your head has a certain similarity to a lettuce plant, thus assuming the IQ you inherited is not very high. Educated mature humans, to whom we all belong, might think of a so-called tree-man when a word like Phytomera is mentioned to them. The commonly known tree-man is suffering from bark-like growths on his hands and feet. He has a peculiar disease caused by a very rare form of immune weakness and a super infection of human Papilloma virus (HPV). The virus "hijacks" the cellular machinery of the person's skin cells, ordering them to produce massive amounts of the substance that cause the tree-like growth known as "cutaneus horns" on hands and feet. Warty roots grow out and disfigure the extremities until they become dysfunctional. Immune-deficiency finally causes the breakdown of the last scarcely present body defense (Bömelburg 2007).

Fauna of Arizona

Unfortunately, the word Phytomera is not describing that rare disease.

The question arises whether a Phytomera exists at all and for what kind of plant/human or plant/beast it stands for. Which characteristics should a Phytomera have? It must contain much water, since humans consist of fifty to eighty percent of water. The water-retaining desert succulent plants adapted to arid climate would fit those criteria. In addition, the water dwelling sea anemones resemble the described conditions. The appearance of a sea anemone is similar to a plant, whereas it is an animal. Maybe sea anemone is closely related to Phytomeras. Perhaps somewhere in this world are Phytomeras. They were never discovered. It might be that none of the experts ever pinpointed their benefits. Thus the miraculous herb, which produces human organs, was never cultivated for its fruits. And so the magic herb still awaits its creation, since we undoubtedly need Phytomeras. They are most efficient and provide the solution to obtain desperately needed organs for transplants. They have many more qualities and are surely the uttermost alternative to stem cell research. The difficult task to get insights in stem cell differentiation can be avoided. The duty would be surrendered to Mother Nature. Whatever scientists are doing, the dream to design organs remains, no matter whether you're wondering if it wouldn't be easier to break into Fort Knox.

The intention of human embryonic stem cell researchers is trying to discover the pathway regulating the differentiation from unspecified stem cells into a variety of specialized cells that form tissues and organs. Scientists orient themselves to the vast amount of research done in veterinary medicine. Should researchers have a closer look into agricultural techniques in order to create a Phytomera? Just imagine if researchers were able to produce such a plant. Would they try to harvest it everywhere? They might become reckless and take every spot to cultivate the desired plant. Without taking any precautions, the natural law and unforeseen environmental misfortunes certainly reduce an abundant harvest. The resulting disaster might resemble the destruction of genetically manipulated plants in the Mid-West of the USA. In 2008, thirty percent of the harvest was destroyed by unexpected heavy rain associated with soil erosion and flooding.

2008 flooding of genetically manipulated crops planted in the Mid-West

The desire for the fruits of the Phytomera would cheat all human intellect, calculations and well-known rules of agricultural planning. Utterly ruthless exploitation of our soils and the narrowing search for profit might cause nature to retaliate. Finally, at the very end researchers have gained nothing. Maybe farmers might remember what Goethe said in such a situation. They eventually join his sentence of self-pity. "Now I'm standing here as a poor fool, still as smart as I have been before."

In general, it would be immensely beneficial and ingenious for mankind when agriculture possesses a plant whose fruits are kidneys, or little hearts or possibly nerve cells. We only have to wait for the harvest. Transplanting those organs will solve all our problems. These plants would have to be highly protected. The death penalty would await anyone who tries to lift the secret or genome of the miraculous plant, regardless of whether the law exists or not in that wonderful country. Blessed is the country that possesses the geographical territory for growing it. With that plant, the citizens automatically gain the status of immortality.

All other countries, which still depend on stem cell research, would have to exploit their women to maintain their egg cell resources. They undoubtedly would fall under the category of a developing third world country. The females of that country would suffer enormously under a psychological and physical disorder due to the massive application of hormones. Nobody could help them. Even the most renowned psychotherapists would refuse to treat them. They could only receive very potent psychopharmacologic drugs and would be thereby condemned to an apathetic life. They would likely fall into a persistent "vegetative state." In addition, they would have to carry the burden of infertility as soon as they reach their twenty-fifth birthday. The inability to pass on life would be simply a side effect of a forced participation in reproduction programs. Generally, established tissue and organ banks would be in greatest need of organ supply, obtained from abortions. Eventually organs would be sold. Finally it would be unavoidable that mothers of killed babies suffer under the post abortion syndrome. Only to exemplary citizens would it be granted to have some children. This privilege would ensure a certain birth rate, to maintain their social welfare system.

Women and girls from poor rural families make up the majority of exploited creatures. They would be forced to undergo medical procedures and to swallow enormous dosages of hormones. It would be the perfect recipe for disaster, because the medical treatments definitely cause serious health problems. Unfortunately, something else would come up. Nobody ever considered that the impact of the high hormone levels would cause females to lose their instincts. Consequently, they would not know how to provide the necessary food supply for their children. Neighbors would be forced to watch over the children of their country, to prevent their starving to death. This is a scenario similar to a clinical case in which a dog was misdiagnosed with pseudo pregnancy, even when she was pregnant. Falsely administrated androgens fully stopped the breeding instincts of the mother dog. The newborn puppies needed to be brought up by humans (clinical case/Veterinary University Hospital, Giessen 1994). The well-being of the sweet little offspring is worth every trouble and time without any question. The dog did abandon her puppies due to hormones and the very abnormal behavior agitates our conscience, even more if we see helpless creatures. To watch something heartbreaking motivates us to believe in the most unbelievable cases.

We establish a much stronger relationship if we are able to see things with our own eyes. Moreover, it is not a question that we can only see what is present. Scientists make human embryonic stem cells visible under the microscope. For them it is a real fact that these cells have the potential to develop and to differentiate into the 220 types of cells found in the human body. Scientists link this potential to the expectation that they will be able to figure out the pathway of differentiation. It is only a question of time until that expectation will be fulfilled. Success is also dependent on certain external factors and surrounding conditions, which have to be present. This potential of the cells produces an expectation, which stimulates the confidence of researchers. Science is certain that something it determines will be established in the near future. You can imagine that a very talented Michelangelo is surely capable of realizing the potential for a beautiful sculpture, as soon as he is sizing a marble block. With no difficulties whatsoever, he would create David out of the raw stone. Wolfgang von Goethe was inspired from a certain sentence he used in Faust. To quote: "Whatever you can do, or about which you dream you could do: start to do it right away. Adventurousness carries genius, power and attraction in itself. Do not wait or hesitate to begin with it." This advice makes someone wonder whether scientists in our time reflect on their attitude about what they are doing. The question to be determined is who or which institution has to justify the proper use of a certain potential? Who makes the decisions that an undertaking does not harm others and serves for a larger benefit for all? Who is going to be in charge to oversee the right proportionality and good intention? Which authority is in charge of dealing with unintended misunderstanding associated with the performance? Has Aristotle indicated in his Material Cause the answer to free us from the controversy associated with a desirable potential, for example the use of human embryonic stem cells? The material something is made from will determine its application. The intended future use depends on the predisposition or natural tendency. Michelangelo possessed the unique expertise and the extraordinary artistic talent to transform an ugly marble block into a marvelous sculpture of David. By doing so, he added a delightful uplifting valuable contribution for the benefit of further generations.

Researchers in our time see an enormous potential in the pluripotency (potential to differentiate into different cell types) of human embryonic stem cells. Even so human life has to be destroyed very early in its development to harvest stem cells. Is the pluripotency of human embryonic stem cells more valuable for humankind than the individual right of life? However, according to data from researchers, twenty more years are needed to find the pathway of embryonic stem cell differentiation. It is only certain that they have the ability to grow into almost any kind of cell. They may hold a key to treatment, or to a cure of serious medical conditions. But, this might not happen in the foreseeable future for a number of reasons. Is it justified under these circumstances to deny an embryo the right to life?

All human beings have the same right to live. When does life start and when does an embryo become human? Are these two different situations? Biology is telling us that the first instant of new life is the moment of conception. Aristotle had the opinion that the embryo will become human sometime after conception, at the time the soul is put into the body. We do not know the exact moment when the soul enters the human being. Does that give us the right to state that in the early stage of development, for example on the fifth day, the embryo is not yet a human being? The law is telling us that we shall not take the life from what is probably a human being. Nobody buries a man if he is only probably dead, because he is also probably still alive (Fagothey a, 1958). To continue, we may not destroy an embryo if it is only probably nonhuman, because it is also probably human. We might feel comfortable with the safest moral way and this would be to treat an embryo as a living human being, with the same right to life as any other person. The profession of a scientist does not exempt or suspend one from moral basic laws. Extreme or grave necessity or serious circumstances might force a collision of the right to live and justify the selection of who might prevail. Indirectly the death of one person is permitted if under that very rare condition one but both cannot be saved in order to choose the lesser of two evils. The priority of selection should be considered for the one who has the better chance of survival. It might be argued that an older human has actually had the chance to live and enjoy his life (Fagothey b, 1958). Can it be justified to take from a human being the chance to live mostly at the beginning of its development? This statement is based on the excuse, that this sacrifice might provide the possibility to cure a fellow sick man in the distant future. Can scientists claim their act and legitimate their means as doing the lesser of two evils? May one argue in such a way, that the purpose of healing numerous diseases is yielding priority over violating another's right to live? What might happen if the expected research results cannot be obtained because they are too utopian? Would one not be extremely disappointed? Could such a failure strike the psyche? Recently the post abortion syndrome was officially recognized. Can we use it as an example to illustrate non-compliance between expectation and the actual situation? We have had other expectations; we wanted to get pregnant when we are done with education or whatever. The way out of the dilemma seems to free us from the disappointment. Sad to say, that the solution of the problem does not obtain relief from the burden. Indeed a much greater depression will follow. Women might face the unwanted never-foreseen consequences of enduring unhappiness (Mahlzahn).

Of course, the post abortion syndrome is much more complex. Still there is a relationship between our acts and our psyche. Expectations have been deceptive and now we are disappointed. The question might be, if someone ever speaks about a so called stem cell syndrome? Researchers might become haunted psychologically by their own acts. Our society makes it very easy to displace responsibility onto a third person. Self-responsibility is the most forgotten virtue. It is much more convenient and we are even trained to shift direct responsibility, because we are not capable of realizing it as our task. It is more common that our own characteristics, desires and acts, especially those that are in conflict with social standards, are gladly suppressed. Carl Gustav Jung, the founder of analytical psychology, describes the confrontation with whatever we are trying to deny as the law of reflection (sometimes known as the law of projection). In general, we are projecting something of ourselves onto someone else. Whenever we have a character trait that we would rather not own, it often shows up in our outer world as events that force us into head on confrontation with whatever we are trying to deny. Thus, we have the choice to master the issue, integrate the lesson into our life or, alternatively, live a life that is full of hypocrisy. The general rule of thumb is that whenever a person reacts very angrily, or launches a moral crusade against behavior in other people, it is because they are angry at the same behavior, or fear of succumbing to that, which lies within themselves (Jung, 1921). In short, I project my behavior, of which I am ashamed, onto someone else. I seek a solution in a defense mechanism to be able to dismiss negative portions of my own personality. Actually, I look for an excuse to spare myself in a compromising situation.

Researchers argue that an embryo in its early stage of life is not aware of destroying him. Can we conclude that the unawareness or unconsciousness automatically is synonymous with a voluntarily offering of being used for research purposes? Researchers consciously denial the value of life with each embryo they select. Might they experience in the short or long run a stress disorder?

The widely read studies of Dr. Phillip Ney are dealing with the awareness of even small children. They experience and realize in a deep way what selection means. Dr. Ney and Dr. Sheridan verified the existence of the so called "Survivor Syndrome".

Children usually know that their mother is pregnant. They also know when she "gets un-pregnant". They are starting to ask themselves, "why was I saved and why were they killed ?" The phenomenon is called Survivor's Syndrome in view of the fact that the behavior is similar to people who survived a disaster. It causes an irrational but real guilt. The small child senses a sibling's arrival and mostly does not welcome it. When the baby suddenly disappears, the frightened child may get a warped sense of his own power to "will people away". Or the child realized that the mother was doing something with the unborn sibling and this is resulting in a fear and mistrust against the mother (Bond 1986). Also, parents unconsciously suffer under side effects of abortion, which is that the acceptance of the violence of killing the unborn lowered a parent's psychic resistance to violence, and abuse the born children as well (Ney 1979).

Are we now completely, totally and unequivocally convinced that it would be wonderful to own a plant that automatically produces organs, which is so desperately needed? Countries that might not possess this plant have to deal with its extinction. Was the intention of stem cell research not exactly the opposite of being in favor of humankind and not to destroy them, at least partially? Stem cell research was seen in serving the well-being of humans with the everlasting goal of having eternal life and not in our downfall. To view stem cell research as the source of destruction of humankind matches with the common sight of Donella Meadows work "the borders of growth", written in 1972 and supplemented by Dennis Meadows in 2004. Both are stating that when the present increase of the world population, industrialization, environmental pollution, food production and exploitation of natural raw materials invariably continues, the absolute growth limits are reached and exceeded in 2100. Almost certainly, this would lead to a rapid and irresistible collapse of the total population and industrial capacity (Meadows 2004). Is it worth striving for eternal life on earth under these unfavorable conditions? Are all research challenges accomplished only to witness the end of the world? The limited resources entrusted to us contain, in particular, the human genome. We have the obligation to care for and recognize our solidarity with our descendants. We have the duty to be anxious about their well-being (NCBQ, 2008).

An expert of the "Dramen über die Erhabenheit", (drama about the sublime) written by Ephraim Lessing, interprets that Lessing felt pain over the life of future generations if we continue our wrong way of life. This concerned him more than any personal insult. Would we risk our life for the benefits of other humans we might not even know, or for those who are not yet born? History is telling us that this is exactly the most beneficial claim of helping humankind (Zepp LaRouche 2001). Pope Benedict beseeches us that we have to find an ethical way to change our way of life (Pope Benedict XVI, 2008). Under this aspect, a Phytomera would help us enormously. Unfortunately, such a plant or word does not exist. The Internet does not mention Phytomeras either.

2 When Science tinkers with food

2.1 Enjoy the Island

The Island of Dr. Moreau (Wells 1896) is an extraordinary, exciting book. A man is stranded on an island only to find out that two scientists have chosen this place as the perfect location for their experiments to clone animal-human beings. Confronted with mixed creatures, part human, part animals like pigs-, dogs- and bear-humans, the island is the worst nightmare for the stranded. Dr. Moreau and his colleague became victims of their own creatures. Afterwards the

Chimeras reconvert into their original species. The story ends with the sheep from the island being infected with Scrapie. Scrapie is a disease known since the 18th century (1732). The name Scrapie describes symptoms because affected animals will compulsively scrape off their fleece against rocks, trees or fences. Scrapie is a fatal, degenerative disease that affects the nervous systems of sheep and goats (Selinka 2008 a). Malformed proteins called prions cause Scrapie. A prion, the smallest known infectious agent, is composed primarily of protein. It is unlike a virus or bacterium and contains no nucleic acid, that is, no DNA or RNA. The very resilient agent is considered to be the cause of various infectious diseases of the nervous system (such as Creutzfeldt-Jakob Disease and Scrapie).

Misfolded prion proteins also cause Bovine Spongiform Encephalopathy (BSE), commonly called Mad Cow Disease. Prion proteins occur naturally. They utilize an important protective function for the nerve cells. Infected individuals are carrying a variant of the normal proteins. That means a naturally occurring protein can convert into a disease-causing form. Transmission can occur when healthy animals are exposed to tainted tissues from other disease carriers. In the infectious state, the native cellular prion proteins deform. An exponential cascade goes on to deform further prion proteins. Prions aggregate in the central nervous system (primarily in the brain) and form plaques known as amyloids. They disrupt the normal tissue structure. This disruption is characterized by "holes" in the tissue with a spongy architecture due to the vacuole formation in the neurons. It is a non-inflammatory pathologic process in the central nervous system with vacuolization of the grey matter (spongiosis). The disease has a long incubation period and lesions are limited to the central nervous system without inflammatory or immunologic reaction. With accumulation of an abnormal form of prion protein death occurs. Remission is unusual, because of the absence of available treatment (Seidel and Kördel 2007 a).

2.1.1 Transmissible Spongiform Encephalopathy

Carriers of the malformed prion proteins belong to the unusual group of progressive, degenerative neurological diseases known as Transmissible Spongiform Encephalopathy (TSE). TSE occurs in animals and humans. None evoke a host immune response.

In humans, sporadic Creutzfeldt-Jakob Disease (CJD) is the most frequent form of the disease under this group (85% of all forms of TSE). The etiology of CJD is unknown. The etiology of other TSE forms is known. Kuru is a type of Transmissible Spongiform Encephalopathy (brain degeneration), associated with the cannibalistic funeral practices of the Fore people. It was discovered in Papua New Guinea (Gajdusek and Zigas 1959). Most scientists believe that a new variant of CJD (vCJD) may be transmitted to humans by eating brain or spinal cord of infectious carcasses. It entered the human food chain before routine inspections of food service operations, which are under the jurisdiction of the Department of Health to reduce the occurrences of food-borne illness. Bovine Spongiform Encephalopathy (BSE) was first reported amongst cattle in the United Kingdom in November, 1986. The source of the BSE outbreak is uncertain. The epidemic could have begun with just one cow sickened by a sporadic disease. It is thought that feeding cattle with meat-and-bone meal from Scrapie-infected sheep could have amplified the prion disease. Additionally it was discussed that a spontaneous mutation was the basic cause of Scrapie. According to top encephalopathy expert Joseph Gibbs, one out of every million cattle naturally develops BSE (Greger 1996).

Normally cattle eat grass, but in modern agribusiness cows are no longer herbivores. To increase meat production, modern industrial cattle farming utilized various commercial feeds. They contain ingredients including antibiotics, hormones, pesticides, and fertilizer. The most common protein supplements are soybeans. However, they do not grow very well in Europe and especially not in Great Britain. Meat and bone meal as protein supplements in cattle feed produced from the cooked leftovers of the slaughtering process as well as from the cadavers of sick and injured animals were widely used. As by-products, they were less expensive. Germany requires a certain temperature steam boiling process for animal by-products. This requirement had been eased in Britain as a measure to keep prices competitive. Besides that procedure, a normally common added solvent was simply dismissed. Was the disease caused by profit-seeking humans? Unscrupulous British production methods left behind all precautionary steps and unintentionally designed the fatal Mad Cow Disease epidemic. While it is not 100 percent certain how BSE may be spread to humans, evidence indicates that humans did acquire vCJD after consuming BSE-contaminated cattle products. By February, 2009, it had killed 164 people in Britain, and forty-two elsewhere. Thus, it became a man-made disease, because humans interfered with nature while lacking the proper knowledge. Horrendous and devastating consequences followed. More than 179,000 cattle have been infected and 4.4 million slaughtered (Wilesmith et al. 1988, 1991, 1992; Taylor et al. 1993, 1994). To contain the disease, the British government took a number of steps, including the institution of a feed ban by prohibiting the use of meat-and-bone meal.

As more scientific information on BSE becomes available, it may be found that the disease might be triggered by a variety of factors. Prions cannot be transmitted via the air, or by touching, or most other forms of casual contact. Yet, they may be transmitted through direct contact with infected tissue, body fluids, or contaminated medical instruments or soil. A potential risk is seen in the bioavailability of prions on contaminated pastures. It was observed that grazing land on which Scrapie-infected sheep feasted continuously infected sheep. Great Britain and South Korea usually buried cattle carcasses and infectious agents remained stable in the soil (Selinka 2008 b).

2.1.2. Hypotheses for BSE Etiology

The British biologist and bio farmer Mark Purdey represents a new theory regarding the etiology for BSE. The eco-detective tried to build up an overall understanding of the bio-mechanisms of environmentally caused diseases by straightforward field observations. His compelling as well as disturbing research is focused to create a paradigm shift in our understanding of the relationship of pollutants to disease and health. His comparative analyses are based on scanning the scientific scenery of regions where BSE suddenly erupted. For him a disproportion of copper and manganese in the animal system itself causes the disease. He established the scientific evidence that the normal prion protein has been shown to bond with copper in the healthy brain. Furthermore, he stated that the malformed version of the native prion is triggered by increased manganese relative to copper. He observed that in those areas of the prion disease clusters of high manganese content were detected, whereas in the same soil a copper deficiency was found (Purdey 2006). In addition, he presumed that chemicals like estrogens and steroids accelerate the manganese absorption into the brain. He examined the occurrence of Creutzfeldt-Jakob Disease and other forms of dementia in body-builders (Parlby 1999). His research points the finger at public unease that reveals unknown effects.

A number of unanswered questions generated more hypotheses. It was observed that veterinary doctors from Great Britain administered bovine pituitary-derived growth hormones to beef cattle. The anabolic family of hormones should increase beef production. Instead, these growth hormones must have been infected with BSE, because the animals and their offspring contracted and manifested the prion disease (MacKenzie 1999). The prion disease was caused by medical treatment. It was transmitted by contact with infected tissue from someone with the disease. The prion agent survives normal disinfection procedures that destroy bacteria and viruses. In former times transmission of CJD has occurred in some medical procedures like corneal transplants. Also, a few people contracted CJD from brain operations, done with instruments that were previously used on a CJD patient. Today, instruments that are known to have been used on people with suspected CJD are destroyed. Painstaking detective work by scientists all over the world followed the zigzag course of the disease. CJD was unwittingly transmitted to other humans through routes including "donated" cadaver human growth hormones. The anterior pituitary gland releases growth hormones secreted by somatotroph cells, called somatotrophin. Children, whose pituitary glands produced little or no native growth hormones developed an abnormally slow growth and short stature. Prior to 1985, before the biotechnology to synthesize recombinant growth hormone was available, human growth hormone was harvested at autopsy from the pituitary glands of human cadavers. France treated growth hormone deficiency in children with BSE contaminated bovine somatotrophin. Later on it was observed that the children contracted the fatal variant of Creutzfeldt-Jakob Disease vCJD (Pollmer and Warmuth 2000). This claim is questionable because the treatment was applied in Rotterdam and Paris in between 1973 and 1987. During that time it was more common that a human growth hormone was utilized, since no synthetic growth hormone existed. However, it was impossible to trace back the origins of the peptide hormone. It was customary to extract it from human cadavers, who suffered beforehand from various infectious diseases, or have had dementia (Patel 1993). The pathological manifestation of the observed vCJD disease resembled more the symptoms of Kuru (Billette et al. 1994). It was not possible to identify whether the disease was caused through human growth hormones, or by administrating tainted animal extractions of the pituitary gland. Theoretically, the last postulation is very questionable because BSE was discovered in 1985.

More than twenty years after identifying BSE a new variant of Creutzfeldt-Jakob Disease was found in young humans. This strongly suggested that CJD is food-borne. Contraction occurs when people eat products contaminated with Mad Cow Disease (Crozet and Lehmann, 2007). For ten years, scientists have known that humans originated an atypical form of Bovine Spongiform Encephalopathy. Both have the same type of prion protein gene mutation. The genetic form in humans is also called genetic CJD (Collinge et al. 1996; Dealler, 1998). TSEs are among the worst diseases yet identified. It has to be considered that the agent appears to bind to alcohol-based disinfectants and resists heating to extremely high temperatures, freezing, burial underground for years and the strongest of chemical and conventional surgical sterilization. Does the island of Dr. Moreau tell us what might happen if we interfere with nature, or is this story only an unconscious metaphor?

2.2 Cultural Revolution in the animal kingdom

Vince Lombardi, a legendary football coach once said: "Winning isn't everything, it is the only thing. We want to get ahead, maybe even be consumed by the desire to compete with anything that moved, and as it is sometimes the case, with some things that didn't." In this context we feel very confident to apply many things to reach any goal we want to achieve. However, we are very ignorant about a comprehensive characterization of, for example, the pharmacokinetics and pharmacodynamics in drugs. Anesthesia is used every day. Even so it is impossible to explain precisely the interaction in humans, or mammalians.

For instance a horse needs dental care, but it does not show any signs of being under narcotic influence. Now five farmers try to calm the horse down. You wonder if the animal is showing all characteristics of paradox sedation. The intended stages of anesthesia are not present at all. Eventually you are convinced that the horse is old. There is nothing else you can do. If the proper narcosis is impossible, the poor equine has to go without dental care. Thus, its fate is nearly sealed. For the horse the chances to consume food are gone, if it is refusing your help. Eventually it will end up at the slaughterhouse. You might find comfort in the knowledge that equine meat is very healthy. Particularly for tuberculosis patients, since this disease is increasing due to lack of awareness and drug resistance. There is no doubt that at least the veterinary doctor has done everything he was capable of. He acted also in compliance with the veterinary state-of-the-art methods.

Donkeys need dental care as well

Surely nobody would accuse a veterinary doctor when a pig, by stretching herself out, accidentally crushes one of her many piglets. The numerous offspring are caused by the veterinarian's hormone application to provoke a super-ovulation (the release of several eggs from the ovary) in the pig. Still it is not the veterinarian's fault that the mother pig does not know how to handle the youngsters. Maybe the term stupid pig was initiated from such an abstruse behavior. At least all the endeavors to produce more offspring and all the assessments in breeding expertise are demolished thanks to the pig.

2.2.1 Breeding management

Humans have practiced selective breeding for thousands of years. Breeding and selection of animals is defined as the

31

selection of the most profitable innovative and quality breeder. The healthiest heifers and the beefiest bull were chosen to start a new herd. Comprehensive breeding plans and management are efforts to intensify the breeding services. Breeding aims in cattle are to increase milk production or the meat quality. The aim ensures the availability of good quality animals for draft purpose and for conserving indigenous breeds. The goal of chicken breeding is good meat and egg production. Today multipurpose warm-blood horses are desired. They have to be easy to train. Therefore quality horses with proven bloodlines are recruited. Graf Johann XVI von Oldenburg (1573-1603) started many breeding farms in his region to produce war-horses. They were originally developed as strong carriage horses. The early Oldenburg horses were well known for their consistency in confirmation, great power and their magnificent coal black color. They were also famed for their kind character and extreme willingness to work under saddle, in front of a carriage or in the fields. Breeding goals have changed. In our times Oldenburg's are used for show jumping, dressage and three-day events as well as occasional driving (Bintz 1995).

Most people choose a dog or cat breed for emotional reasons. They love their physical appearance, behavior and certain characteristics. In addition, other aspects are desired like hypoallergenic animals. For that purpose, Labradoodles are bred. They have the personality of Labrador but will not cause any allergies and thus are great family dogs.

Recent developments in biotechnology applicable to cattle include artificial insemination and embryo transfer of dairy and beef cattle. In farm mammals, early embryos can be removed from the uterus of their donor and transferred to the uterus of other recipients for development to term. Embryo transfer is broadly used in reproductive management of many species. This ensures to rapidly multiply and breed the most qualified genetically superior cattle within a short generation interval. In 1890 for the first time embryo transfer was undertaken in a rabbit. The offspring did not show any genetic influence. The recipient was only completing the gestation period. This outcome confirmed that a surrogate mother could raise another's embryo. In 1930, the first bovine embryo was collected. However, it took until 1951 before the first embryo was successfully transferred. The method provided that a larger number of superior offspring could be produced, in a shorter time as possible with normal gestation. The producer's life span covers usually six to seven generations. Obtaining a super bull from a superior cow was limited over time. By the use of super ovulation and embryo transfer, the rate of improvement was dramatically increased. Today this technology is commonplace.

Dairy producers benefit from availability of worldwide embryo transfer. It also reduces the possibility of catching sexually transmitted diseases. It can also be used to treat infertility, because reproductive traits are lowly heritable. In addition, a rapid genetic change within a small population is possible. Recent developments with embryo transfer include embryo splitting. In an early stage of the development, the cells are bisected to obtain identical twins. The limiting factors are the cost-intensive equipment and the technical expertise. Embryos can be sexed. The method is still imperfect. It is very slow and detrimental to embryo survival. All these procedures make the impact of embryo transfer limitless. Unfortunately, the success depends on many factors, including management and selection of the donor, recipients and offspring. Last but not least, the expertise to collect, handle, and to preserve the techniques of implantation determine the profitability and marketability of embryo transfer (Seidel 1981). In addition, a slight hesitation to manipulate the gene pool through selection will always accompany the procedure. The special selection of up-to-date desirable characteristics will cause the deprivation of a naturally grown biodiversity. The genetic variance is no longer present and inbreeding might occur.

2.2.2 Artificial insemination the first great biotechnology

Today the breeding business is completely based on bovine artificial insemination. Farming cooperatives are specializing in that field. Preservation of bovine spermatozoa by freezing in straws makes it possible to select every bull. Artificial insemination is called one of the first great biotechnologies, applied to improve reproduction and genetics of farm animals. It has had an enormous impact worldwide in many species, particularly in dairy cattle. In 1784, Spallanzani performed the first successful insemination in a dog, which whelped three pups sixty-two days later. Lazzaro Spallanzani (1729-1799) originally trained to be a priest, has had a great interest in natural history. At the age of twenty five, he taught this subject at the University of Pavia. For a very long time artificial insemination, as practiced by bees and other flying insects, has played an important role. The use of artificial insemination in animals is a human invention and has had the most impetus to develop other technologies, like cryopreservation, sexing of sperm, estrous cycle regulation for pregnancy determination and conception rates, embryo harvesting, embryo freezing, culturing and transfer, and cloning. New, highly effective methods of sire evaluation were developed combined with a genetic improvement and the control of venereal diseases. Electronic networks provided a springboard for the exciting developments. A new generation of reproductive physiologists and biotechnologists boosted the discovery and its establishment for a widespread commercial use and its breeding efficiency. With artificial insemination, the elite bulls were not limited to wealthy farmers any more. Certain principles were required. Optimal selection programs were established to transmit superior genetics for milk and beef production. Super ovulation, synchronization of estrous and manipulation of embryos lead to major advances in animal production.

In horses, the semen quality and breeding efficiency are affected by the season. Artificial insemination started in Russia in 1899, because of the need for military horses. World War II ended the use of horses by the military. At that time veterinary schools even discouraged students from pursuing equine careers. Horses were only used in racing and circuses, for ranching and the amusement of the wealthy. The decline in horse population was also due to motorization and increasing mechanization.

Plowing competition in Wisconsin 2010

With the new horse-powered technology, working horses disappeared from rich cities and from the fields of rich countries. Forgotten were the times of the so-called "change from hand power to horses," that characterized the first American agricultural revolution. Today the agricultural need of working horses is only given in developing, third world countries. Research on artificial insemination in horses to improve reproduction was eliminated only when the horse population had declined. Especially after World War II several equine breed organizations inhibited research and application of artificial insemination. China did not have restricted regulations and undertook pioneering efforts to establish this technique in horses. Frozen stallion sperm was never seen as advancement, since breeding was mostly performed during the spring breeding season, or during the fall. Instead, the use of cooled sperm was established, wherein the sperm was inseminated within forty-eight hours after its collection.

Artificial insemination has had an enormous impact worldwide in many species particularly in dairy cattle. The list of artificial inseminations also includes: sheep, goats, pigs, other domestic mammals and endangered mammals. The initiation of artificial insemination and its acceptance by the public was difficult. On one side the opinion existed that research should not include anything that had to do with sex. On the other hand there was a fear that abnormalities in the offspring itself might develop. Thus, these experiments were not supported for a long time. Especially because the opinion was established that artificial insemination would destroy the bull market. Soon the public became better informed. The goals were explained. The ethical approval of the application was granted with the demonstration of positive facts. In the end it provided a remarkable new system of breeding efficiency. The close collaboration between farmers and researchers paid off. Altogether, this helped to dismiss the hesitations and convinced the agricultural community (Foote 2001).

2.3 Frankenfood built from scratch

2.3.1 Image of a clone

The acceptance of artificial insemination technology provided the impetus for developing other technologies. Reproductive research moved quickly on to engage in more challenges and to discover new highly effective methods. The first in vitro-fertilization of a bovine egg was performed in 1983. Breeders realized that it might be possible to preserve valuable genetic material through cloning. Dolly, the second cloned sheep was born on July fifth 1996 in Scotland (Dr. Jan Wilmut in his lecture in Madison WI 2007). The pioneer Jan Wilmut created headlines through his breakthrough reproductive cloning and his much-celebrated sheep Dolly. In general there are two main purposes for cloning: reproduction and medical therapy. The initial stages of both procedures are identical. Cloning technology is an in vitro technology in order to generate a genetic twin of another organism. Dolly was created in a process called "somatic cell nuclear transfer." In vivo, before conception both haploid nuclei (containing one half of the normal number chromosomes) of the sperm and egg cells are merging into a zygote (embryo). An embryo is created as soon as the nucleus becomes diploid (containing the normal number of chromosomes). Scientists remove the nucleus of a haploid egg cell and replace it with genetic material of a diploid adult donor cell. The reconstructed embryo now contains the exact identical DNA from the donor animal. In order to stimulate the cell division, chemicals or electric current are applied. Once the cloned embryo reaches a suitable stage, it is transferred to the uterus of a female host where it continues to develop until birth. Additionally, remarkable success was seen that the genetic material from a specialized adult cell could be reprogrammed to generate an entire new organism. It was previously believed that once a cell specialized into a heart, liver or bone cell it would be a terminal permanent procedure and from that step, a reprogramming would never be possible again.

On day three in most species the in-vitro (outside the body) created embryo has reached his eight-cell stage. In this stage each of the eight cells, if separated, can develop into a full embryo itself. Calling the cells Omni- or totipotent recognizes this capacity. These cells can differentiate into embryonic and extra embryonic cell types.

8 -Blastocystes

8 - cells

Each of the eight cells is omnipotent

At that stage the embryo is brought back into the uterus of the hormonally prepared female host. The embryo will divide many times into a blastocyst. With reaching that stage on day five the implantation of the embryo occurs in an animal specific manner. In monkeys and humans, the embryo implantation happens around seven to nine days after ovulation. This process initiates the invasion of the trophoblast into the uterus and subsequently the decidualization of the uterine endometrium (Breburda et al. 2006 a).

The blastocyst, formed in early embryogenesis, looks like a signet ring. The signet is called embryoblast, from which the embryo develops. The trophoblast can be compared with the ring, which encases the blastocyst cavity. The trophoblast invades in the uterus-endometrium and forms the placenta. The placenta formation and the length of pregnancy depend on the species.

A gene double or a clone can replace high-performance animals. Beautiful talented jumping horses, or fast racing animals can thus from that prospective live-forever. Racing horses are castrated for easier behavior. Geldings lead a much more sociable and stress free life and are not so distracted by other horses. It's no wonder that the outstanding warm blood show jumping champion E.T. sparked the debate over commercial sport horse cloning. E.T. won everything, from national competitions in Germany to the Olympic games in Atlanta, and the World cup in Geneva in 1996. The champion has shown his talent and the performance of his genetics. The only thing the 17-year-old gelding couldn't do was to father foals. His Austrian rider Hugo Simon announced that he agreed to have his horse cloned to save its genes. Cells from the horse's chest were taken, cultivated and frozen in liquid nitrogen at -196° Celsius. On July 17, 2006, a clone of E.T. has been born at College Station, Texas. The foal, named E.T. Cryozootech-Stallion is a genetically identical copy of his sire. He will be used exclusively for breeding purposes (Priehn Küpper 2007). The success rates of some individual steps involved in equine reproduction in horses can be performed. But compared with other species, production is still low with this technique and far away from allowing their use in routine protocols as in cattle. Clinical interest, which allowed the utilization of live mares as donors of in-vivo matured oocytes is not efficient in the horse, thus oocytes are obtained post-mortem. Cloning requires equine ovaries as raw material. Slaughterhouse mares were the primary source of ovaries and oocytes. However, 2007 marked the end of equine slaughterhouses in the United States. Efforts of the Humane Society of the United States shut them down. The closing of the horse processing plants has had profound effect on cloning. Maintaining a herd of mares to supply ovaries is time-consuming and expensive. Equine geneticists negotiated to get ovaries from processing plants in Canada. The equine cloning laboratory ViaGen transferred the team to produce first clones in Alberta. The facility is sixty miles east of a horse processing plant, which provides an ample supply of equine ovaries. "We have nearly forty horses on a waiting list to be cloned next spring", announced one of the scientists. Half of them are from the United States. But horses are really from everywhere: Germany, Belgium, France, Denmark, Egypt, Colombia and Mexico. Even China is interested, but the company is not set up to receive tissue from so far away. Cloned horses will not automatically replicate the successful performance of the original animals. To reproduce a champion is more than genes. Actually genetics play only a fundamental role whereas several other factors, including environment, training and nutrition mainly determine whether the horse will be outstanding. The cloning price tags of $200,000 will not permit the procedure in normal day-to-day business (QuarterHorseNews 2010). Animal cloning raises concerns on numerous levels. Complications include abnormal high health issues in the cloned animal, which have a negative effect on life span.

2.3.2 Scientifically precise high standard slaughter animal

Genetically identical cloned animals are attractive to the industry because ranchers would be able to keep their favorite livestock. Edible products from cloned cattle, pigs, sheep, chicken and goats are milk, eggs, and meat products. How soon will food be generated from cloned animals? Some consumers have urged the FDA to address the moral and ethical concerns of animal cloning before they allow permission for the commercialization of cloned food. If the FDA will give the green light grocery stores will sell food from the offspring of cloned animals. The cloned animal will probably not be slaughtered because of their high price tag. Allowing cows to reproduce "naturally" or by artificial insemination is still cheaper. Especially since scientists from the Roslin Institute in Edinburgh needed 277 cell clones to create Dolly. She was the only animal that matured in the host sheep. Cloned livestock with uniform characteristics will be used only as a breeding animal. This method evokes that the gene pool and its broad variety are going to become narrowed. Selected are only the requested attributes. With the desire to create certain quality exhibiting animals, other breeding traits will decline. Eventually the special characteristics of one particular cattle breed will be extinct. For example, the susceptibility to diseases, or the optimal adaptation to a climate might disappear in favor of a higher milk-fat level. These high performance animals need more cost-intensive and preventive care, such as special vaccinations, treatments against virus and parasitic diseases and so on. Visitors, even family members or the veterinary doctor are becoming suspects for transmitting germs when entering the barn-stables. Disinfection devices would have to be available for the stable personnel. A source of contamination is seen with pasturing the cattle, which exposes them to wild live animals. The only appropriate place for such "super-cattle" would be a stable with special equipment's like ionizing ultra violet (UV) separating sources. The animals need UV-light for the Vitamin D utilization. Without Vitamin D, the proper calcium utilization cannot take place. Pregnant cows are in a special need of calcium. Once the cow calves and has a large calcium demand for milk production, her hormonal system cannot reverse the calcium from the bones fast enough. The consequences are calcium deficiency and milk fever. Vitamin D for three or eight days and UV-light prevent milk fever. Dogs and yard cats are also reservoirs for germs. Every rat, mouse, fly or cockroach is able to become a bacteria catapult. Even the wind could spread infections. It might become questionable whether you are in a stable or in an ultra-sterilized operation room.

Cloning allows passing on of exactly the same genes. However, it depends on more than genes to be equal to the clone donor.

Cloned animals are not as easy to house

The achievement depends on a certain environment, the necessity of training and adequate feeding. The physiological capacity in a cow is reached in a temperature sphere of 50-85° F. The attention should also be directed to the hooves. If they are too long it will cause stress. This is influential in regard to the milk production. The physiological comfort also depends on "personal hygiene." Brushes relieve the animals of most ectoparasites. The times of a single cow or a small number of cows belonging to the family clans are history. A balanced protein and energy diet must be designed. Using averages or "book values" for feeds can result in an over- or under-feeding of certain nutrients. The livestock itself must be adjusted to common treatments. The animal cannot be scared or provoked. Otherwise, the whole cattle herd would break loose in panic. The range of potential environmental impacts is unknown and uncertain in the proliferation of cloned animals. So far it has not been investigated. Cloning raises concerns on numerous levels and poses already serious threats to animal welfare. Engineering these animals for more intensive production is associated with great animal cruelty and suffering. To create cloned animals is resulting in a loss of the biodiversity. It will have significant implications for the environment and the ability of cloned herds to withstand diseases will be almost gone. Cloning seems to be an incredibly inefficient technology. Hundreds of animals suffer as their eggs are harvested, or as they are often repeatedly surgically implanted with embryos in an attempt to produce just one clone. Last, but not least, severe health problems plague cloned animals and claim the lives of most neonates. A commonly observed problem with these animals is the Large Offspring Syndrome. Animals are twice the normal birth weight. One lamb was reported as being five times the normal birth weight. Surrogate mother animals have to undergo an extremely painful and stressful labor and delivery. Often surgical intervention is required to deliver the baby animal. The Large Offspring Syndrome is very often accompanied by visible abnormalities of organ growth, which causes sudden perinatal death. Cloning is pursued to intensify livestock production. The gain is to have animals that grow faster so they can be slaughtered sooner. Also animals should be raised on a smaller space. The rise of "factory-farming" is already accompanied with serious animal health problems. The practice of raising livestock in confinement at high population density lets animals grow so big and quickly that the bones break. They are cornered to spaces so small they cannot even turn around (American Anti-Vivisection Society 2006). In some ways cloning is an extension of assisted reproductive technologies. It can be seen as a radical departure from how animals have traditionally been bred. For centuries farmers used selective breeding in which the desired traits are chosen for reproduction. With "purpose intended" breeding, cattle already began to look like bovine body-builders. Since the animals are not on steroids, selective breeding was applied instead. It illustrates how man is using science to control nature to enhance desirable characteristics in animals. Selective breeding is all about managing sex. Scientists act as producers of sperm-machines. A gene regulates the muscle growth. Modern biotechnologies uses top breeders that have the best genetics for some desired quality and assure that effective genes are passed on. Do we modify animals and make them mere manufacturing machines? Are our animals originated like little Frankenstein's? Why else is food originated from cloned animals called "Frankenfood"!

2.3.3 Ethical debacle involves cloned meat

The Food and Drug Administration concluded that cloned livestock is "virtually indistinguishable" from conventional livestock. The FDA believes that meat and milk from cattle, swine and goat clones is as safe to eat as the food we eat every day. For them there is no ethical dilemma and the FDA will sooner or later approve cloned animals for marketplace. Almost no information about cloned animals is publicly available because researchers claim confidentiality. Studies of cloned meat do not exist, and no long-term studies were ever done to determine the health consequences or consuming food from cloned animals. The animals are bred to grow twice as fast and produce three times more milk. Consumers might not even know how their food was produced. Officials said they do not think special labels are necessary. When people cannot make informed decisions about what they buy and what to feed their families, they are forced to unwittingly make purchases that violate their ethical principles. The question is whether the public will consume food from cloned animals? It might be that many will reject products of cloned animals because of unknown health risks. The German Dietetic Food Law is based on product labeling. The labeling requirements define that the animal origin be specified. In Europe people hesitated to buy meat coming from Great Britain, months and years after the export prohibition for Mad-Cow Disease was lifted. People have been afraid of a fraud label change and were not sure what they have in their grocery-shopping cart. In December 31, 2006, Hematech, a biotechnology company based in Sioux Falls, South Dakota, announced the creation of BSE immune cattle by genetic engineering and cloning technology (Weiss 2007). The European group of bioethics stated that at the present moment no convincing arguments could be given to justify food from cloned animals, or their descendants (Beck 2008). In January, 2008, shortly before FDA wanted to release the permission for products of cloned animals, consumers were opposed to it. They were concerned that cloning could result in food that was unsafe for human consumption. The public suggested the negative impact on human health and well-being may only become apparent in the future. It was argued that if FDA intends the commercialization of cloned animals and it is labeled as "safe", that does not mean it should be produced. Once cloned animal products would become certified it would be a dead end road with no possibility to trace back.

Wisconsin, the dairy rich country of the US, determined that they would not bring milk on the market produced from cloned milk cows. Only some highly specialized biotech companies saw a profit in cloning of highly awarded super cows in order to retain the genetic value. Because of the cloning costs of $20,000 per animal companies can become bankrupt (wiwo 2008). There is only a limited market for cloned animals (Priehn-Küpper 2007).

Should a pet be cloned?

One could think that pet owners want to clone their companion in order to enjoy a newly cloned puppy from their beloved pet, once their favorite pet dies. Particularly as it seems healthier to walk the newly cloned dog rather than visiting the deceased pet in the animal cemetery. Cloning a beloved pet is not inexpensive at all. In January 2009, a pet owner from California spent $150,000 in order to clone his dying dog.

Pharmaceutical companies show a keen interest in obtaining genetically identical animals, to test drugs, or to design organs for transplantations in the future. Only few customers are interested in cloning, because for a long time the European Union has forbidden the production or importation of cloned foods. The EU is also concerned about the high hormone concentration in meat products (wiwo 2008).

Hormone concentrations found in meat products are always an important topic. Slaughterers know that stress caused during transportation of the animals and the slaughter procedure itself causes increased stress hormone concentrations. Sick animals are normally treated with antibiotics not certified for humans. It is only permitted to slaughter them after a certain time. The rule was established in order to prevent side effects like drug resistance in humans who are consuming those animal products. Nevertheless, it is common to find traces of hormones, antibiotics and nitrates, originating from fertilizers used in agriculture. It is impossible to remove those traces of drugs from drinking water. The public is still very concerned that food products might be a source and the key to causing diseases. Organic food products are therefore increasingly popular around the country. Genetically manipulated crops are already suspect. By many, it is considered scandalous that genetic techniques were transferred from plants to animals. The negative attitude is obvious because it is almost impossible to export chicken broilers produced by gene techniques. For instance, the European Union refused to import such chicken broilers. Also South Korea blocked the import of genetically modified US beef. Already in 1970, Dr. Henry Kissinger stated "Control oil and you control the nation: control food and you control the people."

Scientists are aware of the powerful industry demands for genetically engineered products. Researchers are afraid of addressing the topics, especially when their economic welfare is affected. Dr. Engdahl, a well-known journalist, observed that it is common practice to suppress unwanted truths. Furthermore, it is the rule and not a rare exception to punish whistleblowers. Scientists who uncover questionable projects will be slandered and called incompetent (Engdahl 2007). They will lose their jobs if they reveal facts regarding food contamination that concern public health. In fear of profit loss, some products are not removed from of the market (Engdahl 2007).

2.3.4 Designer-meat

In July 2000, a successful deliberate insertion of a foreign gene into a specific gene locus (site) occurred in an animal. Thus, conventional breeding was replaced through genetic engineering. A high-tech way was developed to "breed" desirable traits into livestock. An extra piece of genetic material (DNA) is inserted into the animal genome at the earliest stage of development. Researchers manipulate fertilized eggs and implant them in a surrogate animal mother. The embryo will replicate the DNA splice with the rest of the genetic material. Accordingly, it is manifested in the genetic material of every cell from the individual. However, transgenic animals have not been sold yet, pending the FDA deliberations on how to regulate them. The FDA has meanwhile determined that clones are safe to eat. However, genetically engineered animals are not clones. Their DNA has been altered to produce a desirable characteristic. Whether scientifically altered animals will ever reach the dinner plate is still debated. Whatever the decision will be, it would open the door to many more things to come in industry labs or experimental farms. We might then have environmental trademarked "Enviropigs" producing manure that will not pollute as much, or cows without methane in their flatulence (Ritter 2010). Although, the potential profits and benefits might be enormous, there may be unknown allergic reactions. No one has ever eaten such food. David Edwards of the Biotechnology Industry Association seizes it this way: "For future applications out there, the sky's the limit. If you can imagine it, scientists can try to do it" (Jalonick 2010). Maybe we should not believe everything we think!

Nevertheless, we can expect that we might soon get fast-growing salmon or pork that contains heart-healthy omega-3 fatty acids in the local supermarket. On January 15, 2009, the U.S. Food and Drug Administration decided to categorize genetically engineered farm animals or transgenic animals as an "animal drug." The "drug" is the snippet of DNA. Their products will be sold as food or medicine. The animals will be held to the same requirements as conventional breeds treated with hormones or antibiotics. America's consumers have been eating food from genetically engineered crops, such as corn, soybeans and canola for years. FDA will not label transgenic animals. Even so the consumer acceptance of such food products is still unknown. The benefits of engineered animals are a high-tech way of breeding to produce, for example, easy-to-raise salmon or more nutritious meat. Animals like these would never develop naturally and thus could never be produced by old-fashioned breeding, especially not in a very short time frame. Salmon are naturally slow-growing, and do not grow at all if it is cold. Aqua Bounty developed the fast-growing salmon, by inserting a growth hormone gene. Aqua used a targeted gene insertion and replaced the natural existing with the engineered gene. The natural gene is not active all the time, whereas the replacement gene is. The transgenic fish will attain a weight of one kilogram after a year, compared with a one-year-old normal salmon that will only have a weight of thirty grams. Companies now have the duty of showing that their genetic manipulation is safe to the animal and that any food or animal-feed products derived from the animal are safe to consume, says Siobhan DeLancey the spokesperson of the FDA. Gene-manipulation will be passed on to future generations. Such lasting effects need to be overseen to ensure their safety in order to prevent an escape of the genes to natural populations. In addition, the regulatory process for drugs, including animal drugs, is done behind closed doors stated Gregory Jaffe. Jaffe is one of the lawyers of the consumer advocacy group Center for Science in the Public Interest. Consumers want transparency. Siobhan DeLancy (FDA) is convinced that food engineered for its nutritional enhancement should gain the customers approval. For that selling point, she can imagine that it will be labeled (Adams 2009).

"All of the animals, plants and microbes we used in our food system, or agricultural system are genetically modified in one way or another," argues Bruce Chassy at the University of Illinois at Urbana-Champaigne. The traditional way of changing plants and animals came mostly from selective breeding and hybridization. "Genetic engineering is more precise and predictable," explains McGloughlin, director of the University of California's Biotechnology Research and Education program. According to a 2010 National Academies of Science study, last year more than four-fifths of the soybean, corn and cottage acreage in the United States used genetically engineered crops. Back in the 1990's opponents of genetically engineered crops were "laughed out of the room... and they turned out to be right," says Marion Nestle, a New

York University Professor and expert on food studies and public health. Nestle fears unintended consequences in the environment and food supply. She opposes genetically engineered salmon. Trying to prevent mixing with nongenetically modified populations is a big problem, already "Millions of old farm fish escape, not one or two" (Ritter 2010). The AquaBounty salmon's effect on the environment and the safety of the food concerns a wide range of consumer groups. They worry that the fish will escape and intermingle with the wild salmon population, which is already endangered. The fast growing genetically engineered species consumes more food, which is detrimental to the conventional wild salmon. Ron Stotish, chief executive of AquaBounty assured that his company has several safeguards. These fish will be in grown land based confined pools with a very low potential to escape (Jalonick 2010). Scientists assured that a crossbreeding with wild fish might not be possible, because the genetically engineered eggs of all AquAdvantage salmon will develop as sterile females. They argue that the new breed could ease pressure on wild fish stocks. Over-fishing and pollution are quickly wiping out the native global fish supply. According to a May, 2010 U.N. report, eighty percent of fish stocks worldwide are already fully exploited. Fisheries risk running out of commercially viable catches in 2050. Half of our seafood is farm-raised. It has to be shipped from Chile to the United States, which costs as much as seventy-five cents per pound. The fast growing salmon, called AquAdvantage, is chemically and biologically identical to the salmon we purchase at the local fish store. According to the developer, a Massachusetts biotech firm, biotechnology could help to restore America's domestic fish farming industry. Mr. Greenwood, president and CEO of the Biotechnology Industry Organization, remembers when genetically engineered crops were introduced fourteen years ago. "Critics worried that "Frankenfood" would hurt human health and the environment. Since then farmers have grown corn, soybeans, and other products that are resistant to diseases and tolerant of herbicides. Genetically engineered animals are the next step in food innovation" (Greenwood 2010).

In contrast, skyrocketing feed costs caused ranchers to seek alternatives in order to stay profitable. Big cows emerged as a product of the 1950's and 60's, when farmers were focused on getting more meat. Back in the times feed prices were relatively cheap and grazing land was accessible. Nobody worried about the improved achievements of animal fed. Whereas today beef fed on grassland is a big trend in farm efficiency and definitely has its advantages.

Grassfed beef, a gourment delight

Especially since the opportunity is given that smaller breeds of bovines like miniature Herefords and Angus are available. The half-sized minicows are compact cattle with stocky bodies and smaller frames and relatively tiny appetites. They consume about half that of a full-sized cow yet produce fifty to seventy-five percent of rib-eyes steaks and filets. Minicows are not genetically engineered. Breeders focused on the originally smaller breeds brought from Europe in the US in the 1800s century. From them minicows were originated through selective breeding stated Ron Lemenager, professor of animal Science at Purdue University in West Lafayette, Ind. Mini Herefords with their white faces and rounded auburn-hued bodies weigh 500 to 700 pounds, compared with 1300 pounds or more for their heftier brethren. The smaller animals can be grass fed. Lesser grazing land is needed, they will reach their mature weight faster and their meat can be sold sooner. Research oriented on budget-conscious farmers offered an alternative to inefficient meat production. Even so, farmers who raise mini Jerseys which provide two to three gallons of milk every day have to crouch down on their knees to reach the udders. Nevertheless, according to the Department of Agriculture the numbers of smaller farms has boomed in the recent years by growing to nearly 700,000 in 2007 from 580,000 in 2002 (Huffstutter 2009).

Are all our attempts of assisted reproductive techniques currently in use still convincing when consumers increasingly prefer meat and dairy products from grass-fed mini-cows? Do economic benefits out-compete bioengineered food. From the very beginning cows have been evolved to digest grass. Do we keep in mind that mad cow disease did begin in the feedlot? Are cattle feed manufacturers to blame who unknowingly interfered by using other feeders in industrial agriculture livestock? Still we should not forget food shortages and that development in agricultural technology increased productivity sufficiently to feed billions of people? We see in bioengineered foods a technological progress and we feel that grass-fed cows are a regress, or might it be vice versa? The former chancellor of Germany, Conrad Adenauer, confided: "God limited the intelligence of man, but he did not also limit his stupidity."

3 Genetically manipulated organisms (GMO's)

3.1 Corruption of plant-blueprint for the good reason

Genes, made up of DNA are the basic and physical function units of heredity, the DNA code for a type of protein or for an RNA chain. The nucleus contains the genetic material of an organism. Genetic engineering is the name of a group of techniques applied to manipulate organisms to alter the structures and characteristics of genes directly. The technique has been applied to various industries, mainly in medicine and agriculture. The purpose is the production of new strains of crop plants with beneficial qualities. It is a goal-orientated gene combination in animal and plants to increase the efficiency of the so-called traditional methods of breeding. The most desired genes are selected from different carriers of one species for a "new super organism".

Imagine, for example, genetically engineered chickens. The living being might not look like a chicken any more. The birds are kept alive only by structure. They have no beaks, no feathers, and no feet. Their bone structure is dramatically shrunk to produce more meat. What kind of creation would that be? Science is interfering in the natural order or God's plan. All you are seeing in that chicken is artificial. Do not be surprised if this creature does not look anything like its natural counterpart. This creation you are looking at does not resemble God's intention in any case. He created that ground bird to provide humans with delicious eggs and to make sure that especially young kids go with the cock in the chicken feather bed. Also, to ensure further that you have somebody who will wake you up in the morning. All what you might see in a genetically manipulated organism supports your suspicion. It really seems that nobody needs a cock that is crowing. We all have satellite watches. Besides that, raw eggs may contain the risk of a severe salmonella infection. In modern society feather pillows are associated with allergies. Who on earth would like to deal with it? Modern food resources are dependent on meat and the interest is focused on how to get the most out of one creature. Profit oriented biotechnologists combine only the most efficient genes. That is the whole secret. Moreover, for clarification, with traditional breeding that goal would have never been achieved. I hope you are now somewhat convinced about the great advances in science and technology. By the way, any organism that has been modified by altering one or more genes by recombinant techniques is a genetically modified organism (GMO).

Genetically engineered corn in Wisconsin

The first genetically altered plant was a tobacco plant with resistance to antibiotics. Corn and soybean are the two most common genetically altered food crops. They have been altered primarily to become more resistant to their normal plant enemies. Otherwise, pesticides and herbicides, or both would have to be used in higher dosage. Ostrinia Nubilalis, better known as corn borer is a natural enemy or more specifically the pest of corn. Its damage is enormous. The insect is native to Europe. In 1917 the first one appeared in Massachusetts. The corn borer likes all types of corn, especially sweet corn. First signs are "shotgun" holes in the whorl (the funnel-like new growth). The caterpillars chew tunnels in the stalks or ears of corn. Eventually the plants will fall over. The decrease in growth is combined with the shrinkage of the grain size. All together this causes a huge harvest loss. With the establishment of transgenic corn, a modified genome was created that includes a gene from the Bacillus Thuringiensis ssp. (Bt). As a result, the corn varieties produce their own built in toxin in every cell, which affects the corn borer. But indirectly beneficial predatory insects may result in a casualty loss. The first generation of the genetically modified corn, called Bt-corn is in the agricultural farming business since 1995. In 2008, genetically altered corn was cultivated worldwide on 114 million hectares. Bt-corn and Bt-cotton is planted on 162 million hectares worldwide. It is mainly grown in the United States, Argentina, Canada, Brazil, China and South Africa. In North America sixty million hectares of genetically altered plants (genetically engineered crops) are now grown.

The manufactured corn includes the insecticide resistant Bt-corn and the herbicide-tolerant (Ht). The so-called "stacked gene varieties" exhibit both attitudes. More than 100 patents are accepted on different genetic variants of manipulated plants.

Herbicides cannot differ between crops and weeds. The agriculture business could only use selective herbicides in order not to harm the crops. Whereas, those are not as effective and do not remove all types of weeds. Excessive weed growth let crops compete for sunlight and nutrients that often lead to significant harvest losses. Advantages in weed management are seen with herbicide tolerant genetically modified crops. The easy to spray non-selective herbicides are not harmful to corps.

You might get the message across that cultivating the new genetically changed corn, soy, canola and cotton is important. The main reason why US farmers change to this new farming system is to save 20,000 tons of herbicides annually. Whereas, a decrease in the sensitivity to Bt-toxin has been observed in at least two cotton cultivated areas of the US, namely in Mississippi and Arkansas. The cotton bollworm is more and more resistant to the toxic agent. Farmers claim that other genetically manipulated plants lose their effect on parasites, and insecticide use has increased.

A rapid rise in herbicide tolerant GM crops is increasing the application of toxic herbicides. According to a comprehensive report by the US-Organic Food Center, alarming levels (46 percent) of herbicide use took place between 2007 and 2008. This has led to major environmental concerns regarding safety and ecological risks.

Atrazine is used to stop grassy weeds in major crops. The compound, effective and inexpensive, suits production systems with narrow profit margins, and thus became the most widely used herbicide in agriculture. In the US it is applied to corn crops in Wisconsin and much of the Midwest. The European Union banned atrazine in 2004, because of its persistent groundwater contamination. Several researchers call for a ban of atrazine in the US. They are concerned about possible carcinogenic effects and epidemiological connection to low sperm levels in men (Ackerman 2007).

The use of atrazine is viewed with increasing controversy. It is thought to trigger birth defects and menstrual problems in humans. Atrazine levels in water peaks in the spring and increases risk. Babies conceived in April through July showed a higher percentage of the most puzzling birth defect - gastroschisis. In gastroschisis-infants, the intestines, or sometimes other organs grow outside the body. It is an abdominal wall defect. The incidence of gastroschisis is on the rise. Mothers, who resided less than 20 miles from the site of high surface water contamination with atrazine, gave more births to infants with gastroschisis during a study period 1987-2006. Researchers at the University of Washington (Seattle) matched the cases of live born infants with the disease to U.S. Geological Survey database of agricultural spraying. In 2010, at the Chicago annual meeting of the Society for Maternal-Fetal Medicine researchers unveiled findings that demonstrate the link between the birth defect and the agricultural chemical atrazine (Waller et al. 2010).

With grand genetic experiments unknown consequences may occur. Statements are given that the cultivation of the new genetically altered plants will try to feed the world and so concerns are often despised. Nonetheless most people are not comfortable with the concept of altering the nature of food. Consumer surveys show that more than fifty percent would not eat GMO's. Instead they ponder if genetically modified food is harmful or helpful! Whereas the food industry assumes that the same environmental effects will also occur with the use of normal hybrid breed. Thus, no particular labeling for GMO cultivars is required. Toxicity studies, which are normally performed in connection with the use of pesticides, were never undertaken. Potential health hazards linked to genetically engineered (GE) crops are still unclear. Varieties of Bt insect-resistant corn and cotton are in commercial production; other crops being investigated include cowpeas, sunflowers, soybeans, tomatoes, tobacco, walnut, sugar cane, and rice. Resistance to synthetic herbicides has been genetically engineered into corn, soybeans, cotton, canola, sugar beets, rice, and flax. In several varieties of squash, zucchini and papaya, resistance to viral diseases, has developed. Genetic modification can be used to produce crops that contain higher amounts of vitamins. This genetic innovation might not only damage the soil and landscape environment (Bt-toxin accumulation in plants and soil, nitrate enrichment in the groundwater caused by higher induced nitrogen fertilization, increased water waste). It is also causing the death of many kinds of insects. For example, bees and even other mammalians die through side effects. There is an increasing awareness about threats to pollinators by overuse of pesticides. Bees are considered to be most significant to the reproduction of numerous agricultural crop species and are important in the maintenance of diversity in wild flowers. Wild bees contribute nearly $ 3.07 billion a year in U.S. fruits and vegetable pollination (Losey and Vaughan, 2006). The US has seen an alarming decline in bee populations so dramatic that it eclipses all previous incidences of mass mortality. Beekeepers on the east coast of the United States complain that they have lost more than seventy percent of their stock since late last year, while the west coast has seen a decline of up to sixty percent. Scientists call the mysterious phenomenon "Colony Collapse Disorder" (CCD). It is rapidly turning into a national catastrophe of sorts. "The bee's death is accompanied by a set of symptoms that does not seem to match anything in the literature," said Diana Cox-Foster a member of the CCD working group. Walter Haefeker, the German beekeeping president, speculates that besides a number of other factors, the fact that genetically modified, insect-resistant plants are now used in forty percent of cornfields in the United States could play its part as well (Latsch 2007). The Administrative Affaires Court in Augsburg charged a beekeeper to dispose of his entire yearly honey harvest. In the honey, pollen of genetically manipulated corn was detected. In Germany, it is forbidden to sell honey that contains traits of BT-corn, because genetically manipulated food is not permitted (Bienen-gentechnik 2008). In Europe corn is used in general as animal food. Most varieties of corn do not mature in some parts of European countries, because of the climate.

GMO's are frequently criticized because of the possible risk of unintentional spread of transgenic plants by pollen and seed dispersal. As soon as a contamination in between trans-gene and conventional plants occurs none of the plants can be sold in Germany. Plant diversity and seed saving was the foundations of agricultural sustainability and in a certain context the key to civilization. The fear is justified that genetically modified corn threatens the variety of crops. If one type of plant is lost to disease, a population could depend on the healthy remaining crops. In former times during traditional farming the seed was saved for next year's crops. Agricultural corporations will not allow crop rotation and the planting of legumes to fix nitrogen in the soil off-season. A small group of powerful corporations demand crop uniformity and monoculture with genetically modified crops. Crops suitable to the weather patterns of the local environment such as drought resistant plants are disappearing due to greed and hubris by large-scale companies. Despite understanding the history, science, politics and technology, small farmers are threatened and depend on purchasing patented seed. The philosophy that larger is better forbids small farmers to save their year's crop seed for future use. Might even a hunger problem be predictable when we depend on GMO's? It is thought that famine brought about the collapse the Mayan civilization. They relied solely on a restricted variety of maize until a virus destroyed the one sort of crop (Monet 2009).

Yet it is being released throughout our environment that GMO's have an impact. In 1980 the Supreme Court granted seed companies for the first time the patenting of life forms for commercialization (Diamond v. Chakrabarty). By that act the genetic engineering industry gained control over its own genetically engineered organism and thus over our food chain and subsequently over the planet's own genetic heritage for decades to come. No major insurance company is willing to limit risks, or insure bio-engineered agricultural products. Do they fear unpredictable consequences? (Batalion 2009). "The fact is, it is virtually impossible to even conceive of a testing procedure to assess the health effects of genetically engineered foods when introduced to the food chain, nor is there any valid nutritional or public interest reason for their introduction," stated Richard Lacey, Professor of Food Safety, Leeds University in an Internet GMO forum.

Evidence is given that GMO contaminate the ecological system and influence animal and human health. Antibiotic resistances, neurological, respiratory, gastrointestinal and hematological diseases as well as miscarriages, congenital defects increased in humans and mammals. In France, Austria, Poland, Greece and Hungary the cultivation of the transgenic corn MON 810 is forbidden. It represents a danger for the environment. In Europe, the cultivation of genetically manipulated corn is strongly opposed. In 2008, Germany cultivated 4,000 hectares of Bt-corn. The broad population protests against the "green genetic engineering". In 2007, "Greenpeace Germany" reported about the farmer Gottfried Glöckner from Wölfersheim in the state of Hesse, Germany.

Glöckner, an agricultural engineer participated in field trials and had grown genetically modified corn between 1997 and 2001. All together twelve of his diary cows died after being fed this GM corn and silage. The farmer declared the Bt-corn in their feed as the cause. In 2002 he received compensation of forty thousand Euros by the Feeding Company for the first five dead cows, decreased milk yields and vet bills. In February 2002, the farmer stopped feeding his cattle with the GMO's, but by October the same year an additional seven cows died. At this time, his loss was over one-hundred thousand Euros and he called for a proper investigation. The local German district council in Giessen, the Veterinary Pathology Department in Giessen and the Robert Koch Institute in Berlin as well as the University of Göttingen issued a statement in August 2003 that, "the cause of the incidents referred to could not be determined." A chief suspect for the death of the cows in Hesse is the Bacillus thuringiensis contained in Bt-crops. Studies conducted in Japan in 2003 showed that undigested Bt-toxin (Cry1Ab) is present in calf's stomach, intestine and dung after being feed with Bt-corn. The study was also conducted in pigs. Transgenic DNA and toxin protein fragments were detected which means that both survive digestion and break down much more slowly in vivo than previously assumed by the Feeding Company. Until now no long-term effects of eating Bt-corn has been observed and no toxicological testing on the whole GM corn plant have been conducted. Farmer Glöckner fears that his pastures are contaminated with Bt-toxin, with decomposing manure from his cows. It will leach into the soil where it can bind with the minerals and clay and remain harmful to many organisms. Molecular analysis, carried out by French and Belgian scientists disclose the toxin is not Cry1Ab, instead in ninety-four percent the Cry1Ac gene of Bacillus thuringiensis is the potent immunogenic carrier, that binds to intestinal wall of mice and does cause significant metabolic changes in the gut. It reveals that Bt-transgene bear not only a risk to kill more species of insects, they also contain previously unknown toxicities for other animals and human beings. Greenpeace is therefore demanding an immediate ban on the specific Bt-176 and a full scientific investigation into the death of the cows in Wölfersheim, Hesse. Apart from the investigations into the safety of GM food and feed, Greenpeace urge "public enquiry into the serious abuse of scientific evidence by our government's scientific advisors" (Ho and Bucher 2004).

Is our future food all about the difference between rights and wrongs? GMO provoke many questions and ethical issues regarding what we can do and what we should do. Are scientists playing God? By gene technology a characteristic from a daffodil can be transferred to a rice plant with the purpose of growing rice with a large Vitamin A content. Economic and environmental issues in regard to genetically modified rice are often controversial and disputed. Golden Rice produces pro-vitamin A. It has yet to be commercialized to cure vitamin A deficiency in developing countries. Unfortunately rice consumption may be the leading cause of serious illnesses. The sustainability of rice production is threatened by arsenic. Several environmental research centers in Beijing verify that rice products contain extremely high quantities of cancer causing arsenic (four hundred microgram per kilogram). Such products are partially sold in bio-ecologic grocery shops. With traditional rice farming methods paddy rice can in some circumstances contain high levels of the toxic element arsenic. Also in India and Bangladesh, it is common that farmers flood rice paddies with arsenic-contaminated irrigation water. Under aerobic conditions arsenic is set free by microorganisms. Does the food that sustains half of humanity also increase the risk of cancer? Transgenic rice was created, equipped with a bacterial enzyme arsenate Sadenosyl-methyltransferase. This bioreactor plant would convert the toxic inorganic arsenic into a nontoxic substance. Field studies with the new transgenic rice started already in China (Stone 2008).

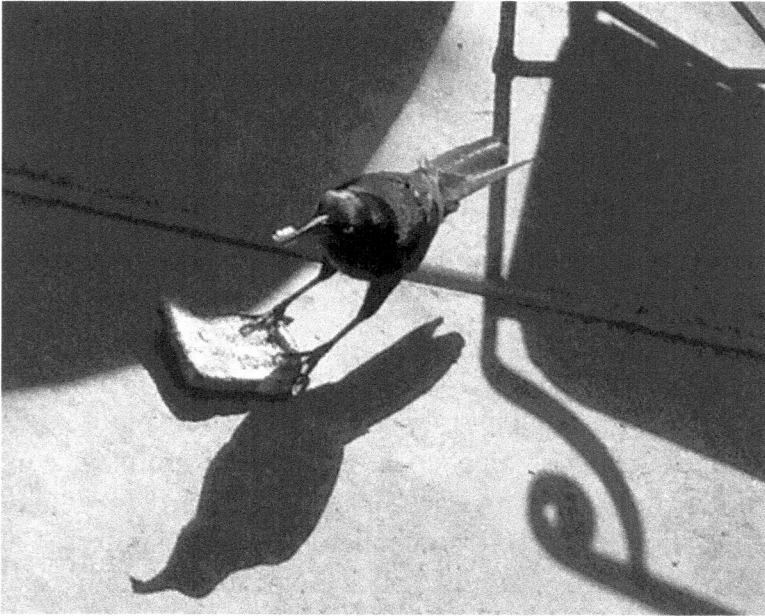

Yet your heavenly Father feeds them. Mt 6,26

Corn is, beside wheat and rice, one of the new world's most important contributions and is adding a lot to the world's diet. Originated in Central America it has been cultivated at least since 3500 B.C. It was the basic food for the Incas, Mayas, Aztecs and native North Americans. In the fifteenth century the first corn plant (Zea Mays) was introduced to Europe. The nutritional value of corn is starch, protein and oil. It is used as popcorn from the whole grains, cornflakes from polenta, or baked goods made from corn flour. In third world countries, cornmeal is frequently used for the production of bread, polenta and pasta. Corn constitutes a major portion of animal (pigs and poultry) feeds. The harvest is divided by its components that are cornstarch, protein, corn-germ, (used for corn germ oil), and fibers like corn hulls (pericarp). Parts of cornstarch are used to produce corn flour. The fibers are utilized for animal food. Cornstarch is the most important ingredient. Potato and wheat corn provide starch too. Cornstarch is used in the paper and chemistry as well as cosmetic pharmaceutical industries. Recently huge corn acreage is being used in regenerating raw materials for the production of subsidized bio-ethanol. Thus, the cultivated fields are exerted for the agrarian alcohol, which is needed for fuels. Thereby, according to estimations of the World Bank, bio-ethanol prompted a thirty to seventy percent cost increase in foodstuffs.

The nation's increased efforts to find alternative fuel resources has unintended environmental consequences. Nitrogen and phosphorous compounds from fertilizer leaked into the Mississippi River and later into the Gulf of Mexico. The agricultural runoff came from the huge corn planting area used to produce corn-powered ethanol. Nutrients and nitrogen stimulate algae growth. The tiny plants will eventually settle to the bottom waters. Its decomposition by bacteria consumes oxygen. Low oxygen water cannot support marine life. Scientists from the National Oceanic and Atmospheric Administration speak about a "dead zone" in the Gulf of Mexico. In 2008, the hypoxic zone covered an area of 8,800 square miles, which is roughly the size of New Jersey said scientist R. Eugene Turner from Louisiana State University. Fish and Shrimp cannot live in hypoxic water. The "dead zone" affects the fishing industry. Fishermen have to go farther out to find living fish, Turner said (Ward 2008). In 2010 the British Petroleum oil spill overlapped some parts of the "dead zone". The oxygen depletion forced many types of fish, shrimp and crab to leave the area or to suffocate. Animals adapted to low oxygen, who live in the sediments, can definitely not survive if the oxygen level falls toward zero. The area of hypoxia covered 7,722 square miles of the bottom of the Gulf and extended far into Texas waters in 2010 (Zabarenko 2010).

Skeptical people always fear that genetic changes put the natural world and the food supply at risk. Indeed in September and October 2000 Safeway and Kraft Foods taco shells were recalled, because they contained small amounts of DNA from genetically engineered corn. A class II recall caused the food chain to replace the item in all of its restaurants. A coalition of consumer and environmental groups tested some taco shells by an independent testing laboratory. It was confirmed that a small amount of DNA or gene for Cry9C contaminated the food products. Cry9C, a gene from a common soil bacterium called Bacillus Thuringiensis, is used to make a protein called pro-toxin which will convert into the active toxin in the stomach of insects to extinguish them. The agricultural company who invented Cry9C Bt-toxin required farmers who bought the seed to sign an agreement to use the corn only for animal feed. Farmers claimed that nobody emphasizes the restrictions but it might be that they just miss written instructions. Whatever the reason might have been some corn got mixed and ended up for human consumption (Brown 2003).

A third of the entire crop harvest in the US produce the protein, Cry9C. GMO opponents fear that genetically manipulated plants given as nourishment to humans abuse them as research subjects. They argue that the very new untested engineered food will probably cause a health and environmental crisis. According to French scientists, genetically modified food poses a risk to human or animal health.

The coleopteran insects Diabrotica Virgifera Virgifera, a very dangerous family of insects, were absent from crops until the late 1990's. It was even forbidden to work with them in laboratories, because they can hardly be eliminated with the known chemical insecticides. The insect might have come from America to Europe through the Balkan war. They were found in the year 2000 around military airports in Italy and France. Monsanto, an agrarian business company seems to have anticipated this problem in developing a transgenic corn. The genetically engineered corn was first introduced in France. To this day, the toxic effects have not been well described. Nothing has been published about the reactions on human cells. A ninety-day study in rats was performed in order to find answers to the question of the expected toxin activity in mammals. Nevertheless, the data we have today are still based predominately on theoretical considerations. Finally, the preliminary data from the experiments taken in some rodents achieved only a bit of knowledge about unexpected effects induced by genetic modification, because the experiments have been directed and interpreted by the Monsanto Company. The outcome of the study is controversially discussed. The main findings regarding kidney and liver as well as blood work abnormalities have been made in favor of the absence of toxic effects. These results are questionable. The interpretations of the data were made by only a few toxicologists. None of them had access to the histological slides or pathology reports of the organ systems. It was claimed that the Monsanto rodent experiment has no scientific basis. It was indicated that the risk assessment on health and environment should be public for GMO's. The minimum could be, like in public research, to repeat the trials to draw clear conclusions from that data. Furthermore the conducted experiments should be expanded and include two generations of rats. An in vivo test would give the final security. It is absolutely necessary to undertake such experiments to test unknown products that cause in vitro negative effects. If those studies are not conducted, the agreement to release corn into the environment for food, feed or cultures, may present a serious risk for human and animal health and the release should be forbidden. Biotechnology would be more easily accepted under the conditions that precise studies have been accomplished. The observation period for GMO feeding studies in rats should be at least two years. The current studies are not long enough to safely exclude long-term adverse effects for humans (Seralini 2005). Greenpeace insists that agrarian companies disclose the facts and to conduct additional research studies (Richard et al. 2007). This is the first and only case in which Greenpeace advocates animal research. The public is also interested in these data. Greenpeace would very much like to see a stop for the permission to cultivate GMO's in the entire European Union. Europe requires labeling transgenic food since 2000. Complex analyses are necessary in order to comply with these regulations. A certain number of people argues that the agricultural industry is more interested in profit than in food safety. The concern exists that in the future only GMO food might be available, because of its superior profitability. To quote Engdahl, soon the animals would have health problems, serious infections and be unable to walk. The result would be the death of the animals. This all may happen sooner than expected (Engdahl 2007).

As a matter of fact, a mysterious illness is causing calves to bleed to death on Bavarian, (Germany) farms since 2007. Veterinarians are stumped over what is causing the deaths: vaccines, genetically modified feed or perhaps even the mothers' first milk (colostrum). "The animals' bodies were covered with drops of blood, and their eyes were bloodshot," stated a Bavarian farmer. In his barn, the first calf bled to death in October 2007. The veterinarian tried everything, he says, including administering vitamins and blood-clotting agents. However, nothing worked, and "within two or three days, they were all dead." The mysterious disease is rampant in Germany's cattle barns. Two-to-three-week-old calves are afflicted. They begin bleeding massively and are often dead within hours. "The disease is still very mysterious, and clarification is urgently needed," says Wolfgang Klee of the Clinic for Ruminates at the University of Munich and Oberschleißheim, Germany. He continues: "Affected animals bleed form various parts of the body, sometimes from skin that is seemingly intact" (Bethge 2009).

Fifty-six cases were submitted to pathological examination in order to allow a better characterization of the disease. The bone marrow was severely damaged. Blood cells are normally formed in the bone marrow, but in the sick animals, bone marrow was literally empty, depleted. Sometimes a bleeding tendency can be associated with bovine viral diarrhea. Whereas bacterial infections or bovine viral diarrhea virus were ruled out. The exact cause remains unknown. A multifactorial cause involving infection, poisoning, immunopathy (any abnormal immune response), or a genetic predisposition is conceivable. As of April 30, 2010, more than 2300 cases of the new disorder have been reported in European Union, 1800 of them are in Germany (Paul-Ehrlich-Institute, 16 July, 2010).

As veterinarians frantically try to find a scientific explanation, clarification is urgently needed. Germany has the highest incidence of the disease, which is called "bovine neonatal pancytopenia" (low blood cell count). It was formerly known as "bleeding calf syndrome". Farmers in this country were advised to stop vaccinating cattle against bovine viral diarrhea. The association of the disease with the vaccination is one of a number of possible causes being investigated. While a causal relationship has not been established the decision to stop selling the vaccination product in Germany is part of a wider strategy to identify the exact cause of the emerging syndrome. UK and other European countries consider the benefits of the vaccine outweigh the risks. Experts suggest that other factors are likely to be involved in the so-called multifactorial disease (Long 2010).

The unexplained illness has farmers deeply worried. The farmers have discussed everything from poisonous ferns to "decades of abusive inbreeding". Chat rooms for women farmers, called "Bäuerinnentreff" are more visited than usual. Postings like: "Perhaps the soy we add to the feed is genetically modified", or solar panels and the "radiation from radio towers" are considered as possible causes of the illness. Or might even a simple vaccination be behind the problem? Some veterinarians suggest: "Do nothing"! "Whatever you do, don't touch them! You will injure the calf the minute you touch it, and it will hemorrhage. Then you won't be able to do anything else for the animal." Until then, farmers will likely have no other choice but to hope for help from above. At a "farmers' pilgrimage" six hundred farmers even prayed to the Black Madonna of Altötting to have mercy on their bleeding calves. The words "Holy Mary, help us in our hour of need," resounded across the pilgrimage place. In the invitation, the organizers wrote: "Because there is apparently no help to be expected from any worldly source, we intend, in making this pilgrimage, to beg for protection and help from the Virgin Mary, the patron saint of Bavaria" (Süddeutsche Zeitung, 2009).

3.2 GMO, the Manna of modern time

Are GMO products a possibility to solve the global food shortage? Have GMO's the potential to increase productivity and would that biotechnological solution be beneficial?

Diseases of plants and parasites cause enormous harvest damage each year.

GMO's, the gold standard

It is still supposed that the mysterious dying of bees is related to pesticides produced or used in corn cultivation. Insecticides are applied to canola and corn in a time in which many plants are blossoming. Bt-toxin is produced in the pollen of transgenic crops. It is very possible that the pollen will be transmitted to the other flowered standing plants. Insects die from ingesting the pollen because Bt-pollen lands on the leaves of the weed plants. The goal is mainly to kill or weaken corn pest insects that otherwise destroy the harvest. It is possible that other species might die as well. Bt-pollen toxin is leading to long-term or cumulative effects on the fitness and survival of insects. The pollen count drops rapidly further away from the field. Still long-term and chronic effects of low-level exposure, as well as pollen flow and how Bt-pollen might affect any kind of insects with the result to extinguish whole populations, are simply unknown. Pollen that produces a genetically engineered protein such as Bt-toxin can also prove disastrous to pollinators such as honeybees, bumblebees (Malone 1998). The impact on beekeepers becomes more and more obvious. Healthy bees are crucial for ensuring the survival of plant species and either herbicides like Clothianidin or Bt-toxin could extinguish these pollinators (Julius Kühn-Institute 2008).

Since 1980 it was observed that a resistance to Bt-toxin occurred in insects. That creates an even higher dependence on chemical pesticides. Studies showed that under selection pressure insects build up the initial resistance to a thousand times greater level in larvae which had not eaten on the treated corn. Even fifteen generations later, a resistance level was found to be around one-hundred seventy times higher in comparison with of the control population (Steinbrecher a, 2002). In general, molecular biology and genetic engineering are still in their infancy. The dangers of providing plants with insecticide-producing genes are not known yet. We just begin to see the environmental impact of these new technologies like: spread of insecticide production in wild relatives by cross-population, the creation of super weeds, the killing of non-targeted organisms, the built up resistance, the damaging effect of the gene products on soil organisms, allergenic or toxic side-effects on mammals, secondary effects on birds and frogs and last but not least, the increase of crop plant diseases as well as altered behavior of the crop plant (Steinbrecher b, 2002). Steinbrecher says in his book "Redesigning life": "Genetically engineered crops made for agribusiness monoculture will lead us further away from sustainable food production and cannot be interpreted as progress". He states that real progress would lie in using our knowledge when working with nature (Steinbrecher, c 2002). The focus of the breeding concept of conventional hybrid corn is to yield a profitable harvest. The naturally occurring parasite defense mechanism was disregarded. During the procedure of sort breeding (means selection of recombinant superior gametes) the natural existing protection against major caterpillar pests (butterfly caterpillar) of crops got lost. The gene responsible for the protection is still present but only growing plants can use the mechanism that extinguishes the insects. The plant natural product biosynthesis of Benzoxazinoid DIMBOA is the selective advantage against the attack of corn pest. In older mature plants, the enzymatic machinery is gone. If it continues to be produced, the plant will lose its stability. Professor Gierl and his team at the Genetic Institute in Weihnstephan, Germany discovered and successfully engineered the genetic complex pathway of the metabolic production to prevent its inactivation in order to maintain the host defense. Further genetic and reverse genetic approaches for enhancing plant disease resistance through natural products are on the way. Thus established by the plant itself, toxic side effects can be attenuated. The only obstacle is that corn varieties, which possess a natural defense mechanism, are less profitable. It is thought of as a recombination of both genes that are carrying the characteristics. This pathway is called "Smart Breeding."

Researchers of the US food company Simplot use only species-specific genes to achieve genetic changes. In the past, transgenic variants from bacteria or even fish were more lucrative. The new option is referred to as "cisgenesis." With it, for example, only genes from the same plant or related plants will be transferred. The goal for the genetic operation is seen in minimizing toxic effects aligned with nutrition. Cisgenic potatoes are changed in such a way that no asparagine's amino acids will be set free in connection with their cooking. Potatoes from such cisgenes are free of the suspected carcinogen acryl amide, which is normally originates from the protein component asparagine's. The Agricultural Research

Center Wageningen in the Netherlands and the University of Zurich are working on methods to reduce apple scarp, a disease in apple trees caused by a fungus called Venturia Inaequalis. The fungus is a cryptic species that causes scab lesions in the apple. The fungus is located in the woody tissues of the host tree. With a cisgenic manipulation, the scab resistance in the apple was induced. Thus, the normal use of copper spray solutions was avoidable. Usually eight annual spraying applications were needed. Copper solutions are applied as fungicides, causing a pollution of soils. Heavy metals are a huge health hazard. Researchers hope that by the use of the green gene-technique the copper toxification of the environment can be solved. Eco-warrior claims that cisgenes should be treated in the same way as trans-genes, because the techniques do not differ. It is possible that artificially transferred genes can change the behavior of cisgenic plants. The position of the gene in the DNA could be responsible for such an alteration.

The question remains whether science is able to change the environment by destroying it. Past times showed us that the improvement of the natural environment by engineering was the policy of the former Soviet Union. In Soviet Central Asia, cotton production was planned to increase dramatically with massive irrigation but led instead to catastrophic salinization (increase in salt content in the soil). Thus, it became a health hazard for humans, animals and plants (J. Breburda 1968).

3.3 Environmental health and protection

Environmental activists, religious organizations, public interest groups, professional associations and other scientists have already raised concerns about GM food. The British World renowned scientist on food safety Dr. Arpad Pusztai, lost his job when he warned about GMO food. He certainly evoked media attention in August 1998. Having said of GM food: "If I had the choice I would not eat it," he continued, "I find it's very unfair to use our fellow citizens as guinea pigs." With this said, it is easy to imagine Pusztai was ideologically opposed to GMO food. But this is not true. Instead he declared: "I'm strictly science based. It is not an ideology for me. I am not a campaigner. I have never belonged to any organization campaigning for or against it." He confessed that he felt he had the duty to speak out, "just to inject some caution into this business," he says. "Make no mistake, this is an irreversible technology. It is no good fifty years later to say: We should have known" (Randerson 2008). He further stated, that he still won't eat genetically engineered food and argued it is insufficiently and very superficial tested. Substances that have a slow acting effect would not be detected in GMO food, because the present regulations do not require long-term safety tests. His research results on lecithin potatoes which were intended for the human food-chain, showed that rats developed immune system defects and stunted growth after a period corresponding to 10 years of human life. Pusztai who worked at the UK's leading food safety research lab, the Rowett Institute, was suspended a few days after warning about GMO food. His statement threatened to damage the then ongoing multimillion PR campaign of the Biotech industry. The Rowett Institute pointed out that Pusztai was old (68), senile and confused. Not only was he suspended, he also did not get permission to speak with the media to defend himself. Later on, twenty-four independent scientists in different countries confirmed the correctness of Pusztai's conclusions. The committee of researchers and physicians further declared that Pusztai was perfectly clear-minded with no signs of confusion or memory defects. The Royal Society in the UK appointed a second review committee that concluded

Pusztai's results were inconclusive and even flawed. Lancet, a world leading scientific journal found the judgment: "a gesture of breathtaking impertinence" (Lancet, Editorial, May 22, p1769, 1999).

The Rowett Institute was formally a charitable and independent institute. But the Thatcher government reduced funding and it has become dependent on the industry for existence. It was revealed that prior to Pusztai's suspending a leading seed company had given the Rowett Research Service a $224,000 grant. Pusztais "mistakes" have been effectively distributed all over the world. People should come to believe that there was no scientific basis for his warning about GE foods. "As a multibillion dollar investment is at stake in the GE food case, drastic actions by the industry to protect theses interests are not surprising. Rather it would be contrary to their responsibility towards the owners not to act with maximum force in such a case," announced the society of physicians and scientists for responsible application of science and technology (PSRAST 2000).

Dan Fagin and Marianne Lavelle wrote in their book <u>Toxic Deception: How the Chemical Industry Manipulates Science, Bends the Law and Endangers your Health,</u> how chemical Industry set up its own scientific institutions to counter health and safety norms in order to keep known health threats profitable on the market.

Is science dysfunctional because it depends on industry? Research has become increasingly expensive. In many States complaints concerning lack of proper financial resources are more often addressed than ever. It is no secret, that even Universities are increasingly funded by industry. In society science has a high status, since it will develop the knowledge, which may help us to understand life. It has the purpose of improving conditions of life on earth. Can scientists continue to be objective, impartial and critical? Opposition of scientists to an important topic is suppressed. Under such circumstances many scientist who are critical to the use of genetic engineering have not dared to express their opinion, fearing that it may threaten their job, their career opportunities, their research grant money or the possibilities getting their research published (ISIS 2001).

Biogenetic engineering appears to have endless possibilities in manipulating plants, animals and humans. Despite the enormous possibilities, its technology can also be applied unethically and recklessly and thus might have the capacity of even destroying us. On August 13, 2008, Prince Charles warned in the Daily Telegraph of the largest environmental disaster of all times caused by genetically altered food. "A gigantic experiment with nature was initiated by the cultivation of genetically manipulated plants. Already today the water balance is endangered in countries like north India and west Australia." In 1990 Pope John Paul II said, "A basic condition for each economic, industrial and scientific progress is seen in the respect for life, in particular of the dignity of a human person."

Our instinct is telling us that we have the responsibility to reduce our selfish behavior that will oppress and abuse others or future generations. The moral and the natural law, which is a law of rationality, greatly restrict our actions. It is widely accepted that it is against the environmental ethic and therefore it is prohibited to destroy things purposelessly.

However, humankind has the privilege in his ability to use nature for the benefit of the humanity. Basic science and modern technology is par excellence employed to exercise our dominion over nature. Science is also accountable and has to justify the use of nature for human purposes. Even though it is not required, research can solve many questions pertaining to humankind. In connection with power, we have to apply intelligence and responsibility over the worldly resources.

In addition, ecological aspects have to be included in our actions. Human beings themselves have priority, before the interests of consumers. The integrity and dignity of an employee cannot be abused in order to take commercial advantage. The gift of life of a fellow man or even animal should be respected. It is not permissible to contaminate the natural habitat by environmental pollution. Resources have to be used in a responsible way, because they belong to everybody and not only to a certain country. The poor who are most often citizens of developing nations are mainly forced to suffer in poverty with its attendant high rates of malnutrition, diseases and mortality as a consequence of our misguided focus and exploitation of the ecological system. In nature, we see an order, a harmony, regularity, a cycle and logic. Are we allowed to interfere one-sidedly with nature out of ignorance? We do not know the consequences when we interfere in one natural cycle, because we are still unaware of the cyclical compounds. The future generations have a right that we provide them as well from the Earth's finite flow of natural resources. Today the poor are suffering and tomorrow our descendants will experience our irresponsible actions or the irreversible damage we have caused. We have a moral responsibility of how we relate to a delicate environment. Uncontrolled destruction of soil, plants, animals or inconsiderate exploitation of nature cannot be justified in the name of the progress, or for the well-being of humankind (J. Breburda 1983).

The Cornwall Declaration of environmental stewardship is a coalition of interfaith individuals which provides a concept that expresses common concerns, beliefs and scientific views on the nature of stewardship. Judeo-Christian religious tradition is contributing significantly to clarify our moral and religious responsibilities in protecting the environment (Richards 2008).

Wildfire destroys Southern California forest

The past millennium brought unprecedented improvements in human health nutrition and lives expectancy. For the new millennium the opportunity exists to build on these advances in order to be beneficial to support a larger earthly population. Coincidentally many are concerned that liberty, science and technology are more a threat to the environment than a blessing to humanity and nature. The unspoken concern is that humans are interpreting the control of nature as empowerment to improve the human condition. But they might also do great harm to each other, to the earth and to other creatures. Thus, the moral necessity of ecological stewardship has become increasingly clear (Beisner et al. 2008).

4 Genetically manipulated humans

4.1 Chimera "smart breeding" for humankind?

We are all concerned when hearing about a beloved friend who suffers from anorexia. It was common to treat this life-threatening eating disorder by a total exclusion of family members. Parents should not interfere, as their child would only get better on her own terms. Until in 1980, the Maudsley Hospital in London believed that family could play an active positive role in overcoming their child's obsession with starvation. In an attempt to break the cycle of the disease the family was allowed to join the patient. The support of them in addition to doctors and psychologists cut down the average of duration of an eating disorder from five to seven years to eighteen months (Hall 2009). It seems the bond of the family, the care and tending to each other's needs remain present in our ever changing society. Love, support, friendship are the main values in a family concept. You might want to go back as far as to Adam and Eve to identify a family, a closely-knit group of related people. As far back in History as we know, the large extended family defined the world of a given individual. It is seen as a social arrangement with tribes, clans and households. One on one interaction is the fundament of socializing. A remarkable in utero study of twins investigated whether socialization is already apparent in utero. Three-dimensional ultrasound studies performed at the University of Padova in Italy demonstrated various types of inter-twins contact starting from the 11th week of gestation. The researchers concluded that the performance of movements is specifically aimed towards the co-twin. Researchers speak about "the social pre-wiring hypothesis" and "an inborn capacity for social behavior." Umberto Castiello, the leading psychologist of the study provided evidence that our need to be social appears as early as the 14th week - until the second trimester (Castiello et al. 2010).

Everyone has the need to feel loved and to have the opportunity to a common interpersonal exchange. Already Aristotle defined "man is by nature a political animal." For him animals and humans are equal in their wishes and desire to socialize (Aristotle c). However, human beings are above animals because of intellect and free will. In the 17th and 18th century Hobbes and Rousseau, the founder of the moral positivism, described the state of nature before society came into existence. In these ancient times humans naturally lacked socialization. The so-called "natural man" lived in solitude, guided only by his feelings. Rationality did not exist. Egoism was not known. Humans did love themselves unconditionally. Their self-love was determined by compassion. Simple or archaic societies were governed in that natural state by instincts and passion in which freedom and equality prevailed. Humans were seen as naturally good. In Hobbes and Rousseau's state of nature general health existed, because the weak were automatically eliminated. Hobbes and Rousseau finally stated, the development of language, science and art destroyed that noble, innocent culture. Consequently, society was the corrupting force and was responsible for the failure of the innocent state of humankind. However, language has in particular separated humanity from animals. If humans living in Rousseau's "State of Nature" did not have language, could they really be called humans? Whatever that "natural man" might be, Rousseau's desire was to come back to that "idealistic nature." Nevertheless, the natural state is a totally illusion and has never existed. Theodor W. Adorno, one of the leading recent philosophers and social critics in Germany after World War II judges assumptions like those of Rousseau and Hobbes: "None of the existing imaginary constructs, that are totally opposed to reality, can be seen as improvements. Quite

to the contrary they deteriorate the philosophy of life"
(Möller). Likewise, it is viewed as controversial to proclaim
scientific speculations as serious research results!
Misrepresented truth will have serious consequences when
applied to humans or society (Hulme 2006). Habermas, a
German philosopher and sociologist, stated that truth-seeking
science is not the point of interest any more. Truth itself is
much devalued. In our days, one is only committed to
technical engineering (Habermas 1968).

On April 1, 2008, Josh Hill published a science article in the
advanced online edition of "Nature Gold". He covered the
farcical news that scientists at Linden lab in San Francisco,
California have successfully created a human-bear-pig
chimera. Hill explained that for the endeavor reprogrammed
somatic cells from human and bear were utilized. The
harvested induced pluripotent stem cells of bear and human
were combined in a cell nucleus. Finally they were injected
into a pig blastocyst and inserted into the uterus of a

all creatures are equal to choose the chimera lifestyle

**Each creature has a right to choose
the chimera lifestyle**

pseudopregnant pig, said the young journalist. Hill continues that Dr. Eli Vance, chief scientist is delighted about that huge success. Soon the researchers hoped to commercialize this technology. Dr. Eli concluded that the biotechnological advancements of her lab helped to fulfill the vision of any human to have the advantage of being able to choose a physical chimera lifestyle. Dr. Eli Vance mentioned that she did not use human embryonic stem cells, because she wanted to avoid ethical obstacles or concerns. "Thus, the decision to use induced pluripotent stem cell (iPS) was made on purpose," Eli explained.

Is the Linden lab seriously interested in ethical research? The writer does not even possess the proper technical knowledge. Did Hill focus euphorically on the sales gimmicks in his article and disregard scientific procedures? He mentioned three genetically different species necessary for the creation of the "men-bear-pig" chimera. The method which Linden lab used is very similar to "smart breeding", a mingling process to vary genotypes in plants. To create a living chimera from such obscure origins is an unattainable goal. The article is dated on April 1, a sign of an April fool's joke and not a technical probability. Nevertheless, the question remains as to what kind of nature might have taken place. Is it a human being, with intellect and free will? Alternatively, an animal or maybe the innocent creature to which Jean Jack Rousseau refers. Even his "back to nature sentence" seems absurd, because the described idealistic nature never existed. The only down to earth issue is the ethical dilemma, which accompanies each chimera. The described technique of Linden's lab is full of ethical controversies. They claimed that they were able to reprogram skin cells from a bear and a human. They bred them back until they became haploid cells, as haploid as egg and sperm cells are. Then the haploid bear and haploid human genomes were fused to a diploid nucleus. The statement to implant a bear-human, now diploid nucleus, into a blastocyst is also misleading. The scientist from Linden lab took an enucleated pig egg cell to load it with the suspicious DNA. The in-vitro fertilized egg was artificially delivered into the uterus of the host pig on day three. The host mother pig is not as described pseudopregnant, because it is prior to the embryo implantation. The uterus is only prepared to enable the then developing blastocyst to "settle down" (implant). Last, but not least the goal of reprogramming cells is to back-breed somatic cells to the pluripotent stage. Researchers are returning adult cells into an embryo-like state without ever creating an embryo. Induced pluripotent cells are supposed to be equal to and behave like their embryonic counterpart.

Embryonic stem cells naturally reside within the inner cell mass (embryoblast) of the blastocyst. The embryoblast differentiates and forms into the embryo. The shell of the blastocyst (trophoblast) differentiates into extra-embryonal tissue, like the placenta (Breburda et al. 2004, 2007, 2010). Embryonic stem cells are extracted from the embryoblast of a blastocyst, whereby the embryo has to be destroyed. "The destruction of discarded embryos made embryonic stem cell research deeply controversial. The question of whether to destroy human embryos for research purposes is not fundamentally a scientific question; it is a moral and civic question about proper uses, ambitions and limits of science" (George and Cohen, 2009).

It is believed that induced pluripotent stem cells are identical to natural pluripotent cells such as embryonic stem cells. To obtain induced pluripotent stem cells (iPS), a non-pluripotent cell, typically a somatic cell is going to be transfected by four (or even two) retroviruses. Reprogramming is achieved through viral vectors. The virus is starting a retroviral-mediated reactivation of four (or two) endogenous pluripotent factors. Finally, the somatic cell is transformed, or back-bred into a pluripotent cell. In June 2007 Harvard University in Boston and the University of California, Los Angeles successfully reprogrammed fibroblast cells of mouse. Unfortunately, one of the four genes used might be prone to form tumors. Scientists consider it therefore necessary to develop new delivery methods. They promised that in time a longer less efficient but seemingly cancer causing free process could be developed. In November 2007, James Thomson at the University of Wisconsin-Madison and Shinya Yamanka at Kyoto University achieved independently the same feat using different cocktails of genes and had success in turning back the clock on human skin cells to create induced pluripotent stem cells (Lubbadeh 2009, a). Still, reprogramming cells via viral vectors make them unsuitable because they could cause insertional mutagenesis and unpredictable genetic dysfunction. The main focus was now to address whether it is possible to generate iPS cells without the use of viral or DNA vectors which deliver the reprogramming proteins. In March 2009 University of Madison scientists found a way to insert the genes into the cells temporarily, using rings of DNA, called plasmids. It is hoped to remove the risk of cancer and other problems (Wisconsin State Journal, March 27, 2009). Several other safer methods of making iPS cells will likely be announced, as different scientists try different strategies. In 2009, the February issue of the Journal Cell published an article in which a single transcription factor converted an adult mouse neural stem cell into embryonic-like stem cells. Transcription factors are genes that control the activity of other genes. Hans Schöler's team from the Max Planck Institute for Molecular Biomedicine in Münster, Germany, demonstrated that pluripotent cells could be made without the risk of producing cancer. The specific genes carried by cancer-causing viruses which are normally used to genomically alter the cells, were simply removed. Nevertheless, the extraction of human adult brain stem cells is not very practical and difficult which limits them as a source for research. In April 2009 Sheng Ding in La Jolla, California, has led his group to succeed in a method of a direct delivery of the four reprogramming proteins (Oct4, Sox2, Klf4, and c-Myc). The reprogramming factors were fused with cell-penetrating proteins and channeled into a mature cell to turn it into an embryonic-like state. The protein induced pluripotent stem cells eliminate the potential risk associated with the use of viral DNA transfections. Simultaneously some researchers report that the proteins are too big and cannot penetrate the cell matrix (Lubbadeh 2009).

The establishment of these protein induced pluripotent stem cell colonies took about 8 weeks, which is double the time seen with viral transduction. Also right now the efficiency is significantly low with only about 0.001 % of input cells with protein-based protocols compared to virus-based protocols with an efficiency of 0.01% of input cells (Kim et al. 2009). Researchers suggest that the forming of iPS cells is extremely time consuming because the method is very slow and inefficient with most cells failing to be reprogrammed. From one million skin cells only 3 stem cell colonies can be generated. It also remains unclear whether theses induced pluripotent stem cells can be produced from skin cells of aging patients. It is necessary to overcome the limitations of low reprogramming efficiency and genomic alterations due to viral integration to reach the eventual goal of clinical application.

Scientists blame a kind of epigenetic markers, which they call epigenetic memory to suppress, even inhibit the reprogramming process. The DNA is the blueprint of a living organism. Epigenetic markers in form of chemical tags, called methyl groups are responsible to switch genes on or off. All cells of one organism share the same DNA. It is assumed that the DNA methylation, or the epigenetic factors direct the body cell to different forms and functions, and to develop into liver cell or a kidney or nerve cell. Researchers speak of an epigenetic memory. Adult, or differentiated cells do not need the "stem cell genes" any more. The main hindrance is the removal of the molecular tags which means the methyl groups from specific regions of cellular DNA. "The obstacle of DNA demethylation (removal of methyl group) is critically important in reprogramming adult cells to function more like their stem cell predecessors. It has eluded us for decades," noticed Helene Blau, PhD, and Professor of Stanford's Institute for Stem Cell Biology and Regenerative Medicine. Blau and her research team fused a mouse embryonic stem cell with human adult skin cells to create cells that include DNA from both species to be able to study directed nuclear reprogramming (Bhutani et al. 2010). However much uncharted territory remains in the field of stem cell biology and development.

DNA methylation during cellular development might be explained by the example of a rechargeable lithium-ion battery, we all know from our laptop and cell phones. We all ask us once a while why our batteries lose their capacity. "One of the two electrodes is graphite, a form of pure carbon consisting of sheets of carbon and atoms," explained Dr. Robert Hamers, Head of the UW-Madison chemistry department to the Wisconsin State Journal on May 14, 2010. "Lithium ions are forced between the carbon sheets when the battery is charged and come back out again when the battery is discharged. The first few times that a battery is charged, a chemical reaction occurs on the graphite, forming a thin protective layer. Over time this layer slowly becomes thicker and eventually begins to block the flow of electrical current needed to charge the battery, degrading the performance," concludes the expert. In regard to our topic, methylation occurs as embryonic cells develop and mature. The developmental genes become overly methylated. IPS cells differ from embryonic stem cells in important aspects that pluripotency genes, or the epigenetic memory are difficult to turn on again, or are even permanently inaccessible, due to hypermethylation. "The DNA demethylation in the process of reprogramming somatic cell nuclei constitutes a bottleneck and is especially difficult to expunge," argued Dr. Helen Blau. Scientists are focusing on several molecular hurdles that impede cellular reprogramming. But the most intense research interest right now is to circumvent methylation without disabling it permanently.

Once the best approach is identified, scientists will have the same hurdles and hopes with iPS cells as with embryonic stem cells: figuring out how to grow them into heart, brain or pancreas and other cell types in a way they can repair or replace tissues damaged by disease without harming patients (James Thompson, Wisconsin State Journal, March 27, 2009). Theoretically, iPS cells are identical to human embryonic stem cells, because they are also promoting pluripotency. Induced pluripotent stem cells have yet to prove that they are a safe and suitable substitute for the deficient cells they might eventually replace in a patient. However, induced pluripotent stem cells differ from their natural pluripotent embryonic counterpart. An experiment compared mouse cloning with nucleus donor cells of mice embryonic stem cell to the nuclear transfer of induced pluripotent stem cells. The created embryos were then transferred into the wombs of surrogate mice. In mice, the embryonic stem cells nucleus can generate a new mouse clone. Whereas a induced pluripotent stem cells nucleus only develop to midgestation and will be aborted soon after reaching that stage. It seems that some developmental cues may be missing, states Eggan a Harvard Professor from

Boston, who generated the first patient specific cells from induced pluripotent stem cells (Park 2009).

After all, many scientific unanswered questions remain. For people afflicted with debilitating diseases, embryonic stem cells are holding the promises of hope and a better life. Researchers only lack the understanding of how molecular mechanisms control pluripotency in embryonic stem cells. Unfortunately, this knowledge is of central importance in medicine and science. So far, various hypotheses are generated. Certainly, there is a long way to go before human embryonic stem cell transplants can be used for therapies. The key lies in finding just the right recipe of growth factors and nutrients to trigger the pluripotency of stem cells to become a heart cell, a neuron, an insulin-making cell and so on. Researchers know it would take decades until new therapies might come from stem cells. Nevertheless, researchers need to know the natural pathway of differentiation to imitate it for their purpose.

What were the motives of the Linden lab which create the alleged human-bear-pig chimera? Induced pluripotent stem cells and reprogramming is pointed out as an alternative to embryonic stem cells. Josh Hill the journalist only realized in his article the potential of modern biology which is able to reprogram adult cells to iPS cells and to differentiate them into gametic (egg and sperm) cell's. Reports in the British press indicated that researchers of Newcastle University in England claim to have transformed stem cells into human sperm cells. The possibility of generating gametes from other adult cells raises a host of questions. Current British law prohibits using any cells for reproductive purposes, whereas it is not even clear if the cells are really functional sperm cells. For many it would be unacceptable to produce human iPS-derived germ cells that constitute an embryo (Moreno 2009).

4. 2 Desire to manipulate life

To be able to create life is very desirable for man. From Genesis 3, 2-5 we know the small talk between the serpent and Eve: "God said we're allowed to eat the fruit from any tree in the garden, except the tree in the middle of the garden. You must never eat it or touch it. If you do, you will die." "You certainly will not die!" the serpent told the woman. "God knows that when you eat it your eyes will be opened. You will be like God." Stem cell researchers just love to compare the paradise tree of life with their research. In their view, the fruits are stem cells! The immature fruits are like undifferentiated human embryonic stem cells and the more mature are already differentiated. With time, science will gain knowledge. Researchers are confident that they will obtain mature cells. Their research will make it possible, "that one day we will gain eternal life by consuming those fruits" (First Annual Stem Cell Meeting 2006 in Madison, Wisconsin, USA). Unfortunately, it escapes researchers to realize what happened to Adam and Eve after consuming the fruits. To update the product information for the fruits, the first parents lost eternal life. More precisely by Eve's actions death came into the world. Nowadays it is still very lucrative to be like God and create human life. However, it is worth mentioning that researchers know the Bible. Certainly, they know Genesis verse 1, 28 as well. To quote: "God blessed them and said to them, be fruitful and increase in number; fill the earth and subdue it." The assignment of God is clear. He as the creator of all life asked humans to cooperate with him in giving life to a new human person. In order to remain authentic the mandate is thought to be accomplished, as we all know, by other means than in a science lab. To be politically correct, co-creation is the fitting duty for husband and wife. The proof is obvious, because God created Eve out of the rib of Adam. He was certainly motivated by the thought that it is not good that man is alone. Acknowledging Eve, Adam stated: "That is why man leaves his father and mother and clings to his wife, and the two of them become one body" (Gen. 2, 24). "In this way, humans are about to give and welcome life in an atmosphere of love. It is much more than to execute a physiological reproductive instinct" (Pope John Paul II, 1983). Is it possible to participate in the divine order of creation without God? Or is that clearly a violation of his rules and almost anti-human? God created man and commanded him to bring other people into existence! This is a moral imperative (Trujillo, 1995).

Do scientists who are manipulating life severely clash with the moral imperative? Science always was and still is a product of human reason. Does that consequently mean that research has the permission to execute our possibilities only by the justification that we are living in a new era of biomedicine? It is questionable what kind of life researchers create. The intention of the scientists is to create with the goal of destruction of embryos. A laboratory embryo will unavoidably be destroyed when the blastocyst stage is reached. On day five, the embryo has outlived its usefulness. The act of creation performed in a laboratory is reduced to a biological process, namely the fusion of egg and sperm cells. Researchers doubt that the lab creation is a personal human being from the moment of conception. If you stick to the rules, Adam conveys to us in Genesis that the "*in vivo*" participation in creation always results in a human being. Right from the beginning, the embryo has its own dignity, identity and so on. Nobody would ever question that. Is an "*in vitro*" creation undermining this? Scientists seem to have the attitude to refuse embryos the right to be humans. Does the humanity of an embryo depend on the way in which he was created, from whom and in which location?

The beginning of human life is still debated. A main contribution for the debate is the so-called identity argument. The identity is not immutable, only the temporal expression of that personhood. An old man has the same identity as he has had as a teenager. This reasoning is used in science. We use the same identity statement between two types of terms. For example, water is identical with H_2O.

Water flows, life grows

The dignity accompanied with the identity is the same. The identity of each person is given by its creation. The dignity is also given to everybody who is human. You are not becoming human when you are in a certain embryonic developmental stage. From the moment of conception, cells are dividing continuously. There is no specific point where embryonic development would stop and suddenly change into a different moral status. The philosophical argument of continuation concludes that an embryo must have the same dignity as an adult (Hornbergs-Schwetzel 2008). However, the question of when life begins occupies one's mind and even biology is very clear about that. Some scientists found and are absolutely convinced that human life in all its dignity starts after daily work.

4.3 Stem cells pioneers

In the search for cures for currently incurable diseases, human embryonic stem cell research is expected to be notably beneficial but also extraordinarily challenging. Embryonic stem cells are a remarkably versatile class of cells extracted from an embryo that have the potential to turn into any of the body's 220 tissue types. The more restricted adult cells taken from mature organs or skin are limited to becoming only specific types of tissue.

Scientists have raised hopes about achieving the production of whole organs out of either embryonic or adult stem cells. At least tissue fabricated from these cells should be available. Today the field encompasses far more than just embryonic and adult stem cells: it has expanded into a broader field of regenerative medicine. Scientists consider not only the replacement, but also the repair of damaged tissue. In 2008 at the International Congress for Genetics in Berlin, Eric Lander of the Massachusetts Institute of Technology predicted dramatic breakthroughs in somatic gene therapy in the coming five to ten years. Landers research was decisive for decoding of the human genome. In his opinion, diseases like cancer and Morbus-Crohn are not far away from being cured. Lander promised that we surely might overcome the very unfortunate momentary setbacks and time delays. "Currently we are still scraping the surface". He reassured that many people are fascinated by the possibility of therapeutic intervention in the deep structure of the human genome (Müller-Schmidt 2008). Such treatments appear to be at least several years away.

Researchers marvel about a biological revolution. For researchers embryonic stem cells remain the gold standard for any treatment. This fascination of scientists holds deep controversies and does not find the acceptance in the broader society. The ethical debate will definitely increase with genetic manipulation.

In 2001 British scientists appealed to obtain permission for human embryonic stem cell research. The researchers argued that they do not want to lose their competitive edge in this, "most important and promising biotechnological area." At the same time a very dismissive open letter from Christian, Islamic and Jewish clergyman as well as representatives of the Sikh religion demand restraint from that kind of research. In 2001 Members of the House of Lords, Britain's second parliamentary chamber, supported government proposals to permit research using human embryonic stem cells. During a passionate debate, Lord Alton questioned the morality of treating the human embryo as "just another accessory to be created, bartered, frozen, or destroyed." He further proclaimed: "These are not trivial questions that preoccupy a few moral theologians. They are at the heart of our humanity." Vice Health Minister Lord Hunt is restraining from a further delay of the decision. For him the sacrifice to destroy embryos and to refuse them life is a very necessary oblation, with the purpose of accomplishing vital research. Hunt continued: "The respect we show to embryos should be seen in relation to the promised cure for millions of people living with devastating illnesses and the millions who have yet to show signs of them." Hunt concluded that we have to pay tribute to the sick people as well. "Only embryonic stem cell research could alleviate the suffering for thousands. This type of research is a beacon of hope on the horizon." Between 300,000 and half a million human embryos have been destroyed or experimented on in between 1990 and 2001. Hunt apologizes that there have been no cures, but he is convinced of our willingness to continue to walk a road that has paved the way for more and more demand (Mayor 2001). Since 1990, it is possible in England to perform experiments with embryos up to fourteen days old. They can be used only for very narrowly defined research purposes relating mainly to reproduction. Otherwise, the donated embryos from in vitro fertilization clinics are disposed of biological waste. In England and USA surplus embryos are created, more than needed for in vitro fertilization. With the new regulations, England allowed the Human Fertilization and Embryology Authority to license a wider range of research, including the development of new treatment approaches to serious medical conditions such as Parkinson's disease.

The enormous debate from 2001 promised to accept the given limits of research and to restrain from cloning embryos. The purpose of therapeutic cloning is to obtain patient specific stem cells from five-day-old embryos. That means the cloning method is used to generate stem cells from afflicted patients who are donating themselves a cell. Patients with kidney failure could theoretically donate themselves a healthy new kidney without the fear of rejection. However to grow a new population of a patients' own cells proved practically impossible. Nevertheless, the new law provided the harvesting of stem cells from aborted children (Rötzer 2001).

Great Britain's debate can be regarded as the turning point in legalizing stem cell research. In 2009, supporters of human embryonic stem cell research still argued that the use of embryos is essential to further progress in the field. They suggested embryonic stem cells are more flexible and hold a greater potential for medical treatments in humans. This argument might lay a burden on specialists and put them under constant pressure to authorize increasing research into embryos. Thus, even the recent proposal for animal-human hybrids was somehow predictable. Already in 2001, Lord Hunt was afraid of alternatives. He stated: "The 1990 Act already provides the answer to the question of what happens if and when research into adult cells overtakes research using embryos: embryonic research would have to stop because the use of embryos would no longer be necessary for that research." From the very beginning, Lord Alton proclaimed that we have considerable arguments against the use of embryos in our space. He announced that even without the complex moral arguments, or without being an obscurantist religious believer there are reasons to prefer adult cells. In 2008, Alton stated that for at least 70 maladies adult stem cells are providing real treatments for humans. He declared that adult stem cells offer just as much promise for the medical technologies of the future as embryonic stem cells, with none of the special technical difficulties and immunological complications presented by the use of human embryonic stem cells. He is convinced that a considerable majority of peers see research on human embryonic stem cells as truly insupportable. He does not understand why up until now vast of sums of taxpayer's money, time and energy is spent on human embryonic stem cell research. Adult stem cells are already proven effective (Lord Alton 2008).

By now we should have been very successful and already utilize approved therapies based on human embryonic stem cells. Claims that such research "might", "could" or "may" lead to treatments for many diseases have been made for almost a decade. We still have not achieved any benefits! We are not lacking funds and proponents. All research has been greatly and fully supported. The United Kingdom Stem Cell Foundation, a registered charity is directly funding innovative UK clinical projects with the goal to bridge the gap between pioneering research and proven clinical applications. Thus, stem cell research was uniquely financially established to become world class. Great Britain established by progressive legislation to move research to the most prestigious location. Finally, it is the well-known pioneering stem cell research country. One might wonder why real breakthroughs are not recorded so far. Rather, on the opposite side, researchers feed the world with hopes and promises. They need a great deal of time to prevail. They certainly expect to reach their goal to be able to use embryonic stem cells for certified therapies. Actually so far, they do not have any therapy at all. This is very disappointing, considering the circumstances of the supportive ambience. It is disappointing that by now a small prospering discovery as well as treatments have not found their way into the clinic.

Companies in the business of immortal human embryonic stem cells like US-Biotech-Enterprise or every equal financial online investment service try to attract clients. On one side, partakers are advised about enormous future advantages. On the other side, warnings are made about financial disadvantages when participating in an insecure and very risky investment. What sounds so precarious? Laws were changed. Everything possible was done to support and maintain, as researchers think, the powerful research tools. For the first time England's scientists are permitted to create chimeras. A chimera is a hybrid creature that is part human, part animal. By that step, scientists have begun blurring the line between human and animal. Injecting animal cells or DNA into human embryos or human cells into animal eggs generates chimeras. Researchers are allowed to keep the chimeras alive until day fourteen.

The first chimera experiments were done in the late 20th century. In 1927, the Russian newspaper Russkoye Vremya published the shocking experiments of the Soviet geneticist Ivanov. He allegedly tried artificial insemination of human and ape females with the other species' sperm. The experiments were done at his breeding ape farm Suchumi at the Black Sea. Similar human experiments done in the Third Reich reminded people of Ivanov's attempts. These trials went down in history as horrendous Nazi crimes and still stigmatize Germans.

England allowed some scientists to experiment on chimeras even before the law was established in 2008. Egg cells donated from cattle were enucleated and replaced by a human diploid cell nucleus. In 2003 Chinese scientists at the Shanghai Second Medical University successfully incubated rabbit eggs with a skin cell nucleus taken from a human patient. The embryos were reportedly the first successfully created human-animal chimeras. They were allowed to develop for several days in a laboratory dish before the scientists destroyed the embryos to harvest their stem cells. Jan Wilmut and many scientists refer to China as the country in which the government is most receptive to stem cell research. "China did find the outermost suitable egg vector and a splendid egg supply for the human genome," stated the director of the North East England Stem Cell institute in a lecture held in Wisconsin/USA in 2007. Using the hybrid embryo procedure empowered Wilmut's Research Center to speed up the production of embryos. It is still difficult to receive donated human egg cells. Who can resist getting the desperately needed egg supply from the slaughterhouse? More than 200 egg cells can be harvested on one day, because it is possible to extract them up to four hours after the slaughter procedure. Whereas, the donation of eight to ten human egg cells can be expected in one month. Chimeras are allowed to develop until day fourteen even though the embryos stop growing after two to three days. Beside the law prohibiting implantation of chimeras in a woman's womb, science is not even technically able to support a longer development than day three.

Opponents of chimeras clearly state that embryos with a preponderance of human genes have to be regarded as human beings (Rötzer 2008).

With unlimited quantities of patient somatic cells and egg cell supply from animals, it is now possible to study the development of diseases in the Petri dish. Nobody knows when it will happen that researchers might have the insights required regarding how to turn human embryonic stem cells into any of the cells that the body might need to repaired or replaced. Not everybody is determined and confident. Doubts exist that a patient will actually benefit from human embryonic stem cell replacements. Their critic is not based on being put out of this field by ideological views but science itself set the limits. Stem cells from animal-human chimeras are polluted. Dozens of embryonic stem cell lines already in existence are proved not to be viable. Chimera may only share the possibility of studying how to treat diseases. So far the treatment of diseases is lacking because nobody knows the cause and how and why the disease developed. The common way for science to study treatments for the disease was to wait until a patient appeared in the office of the physician with symptoms. The cause could be long gone by then. The medical personnel only got to see the end stage disease. The Petri dish filled with stem cells may provide insight, because the major steps in the disease process are exposed. Each cell is a potential target for new drugs to treat what goes wrong.

Pluripotent stem cells, induced pluripotent stem cells, natural embryonic stem cells, or stem cells isolated from chimeras can be used for cell culture experiments. Researchers can view them as a potential therapeutic opportunity to examine the impact of pharmaceutics. Inconvenient animal tests can be replaced. The human embryonic stem cell is compensating for the so familiar lab-rodents. Certainly it is difficult to establish pure cell cultures. Concerns are raised about which elements of the culture are human and which are animal. Undesirable tissue components have severe side effects in later transplantations, such that all of these are faraway events in an arbitrarily distant future. Furthermore, it is a very delicate process to generate spontaneous growth in embryonic stem cells. Researchers admit that they are still in the infancy of the basic research. A long time will pass by until stem cells are stable and versatile enough for transplantation into patients (Adjaye 2008).

The current available lines of human embryonic stem cells are contaminated with non-human material. In order to culture stem cells in a dish it is necessary to use a nutrition layer for attachment. The connective tissue is called "feeder-layers." In the traditional culture, human embryonic stem cells grow together with animal-derived connective tissue like mouse or fetal calf serum. In some cases, the feeder layer cells or particles may proliferate into stem cells. Thus, stem cells grown under these culture conditions appear to human antibodies like foreign particles, because of the incorporation of non-human animal material at the cell surface. Stem cells under attack compromise their therapeutic use for humans. Stem cells isolated from chimera embryos unavoidably deal with immunological interferences. Researchers seldom address that topic. Contamination is naturally occurring between both species and has to be regarded as a safety issue. Contamination of stem cell lines has shifted the attention of scientists, politicians and many more toward legalizing and funding the creation of new unpolluted stem cell lines. They did not want to be responsible for the promise of breakthrough therapies for conditions like Diabetes, Parkinson, Alzheimer and so forth that cannot be realized, because of a lack of funds or the use of old cell-lines. Scientists requested starting over with new created human embryonic stem cell lines. Meanwhile a new commercial serum is available completely free from animal feeder layers. Legalization of new lines was connected with competitions. It was feared that leading scientists might move abroad to Britain or Singapore. Others who stayed behind might shift their attention to the, "less versatile adult stem cells". It is argued that the remarkable versatility of human embryonic stem cells has to be studied in order to understand the secrets of adult stem cells. Thus later on adult stem cells can be used as alternatives (Mlynek 2008).

In 2008 the vice president of the German Research Society (DFG) was convinced that human embryonic stem cells are truly indispensable. For him induced pluripotent or adult stem cells might not prove as stable and as versatile as embryonic stem cells when they are transplanted into patients. Human embryonic stem cells are uniquely qualified as research tools as exemplifying the gold standard in modern medicine explained the German Research Society (DFG) vice president (Hacker 2008). In 2008, the German law was changed in order to "work with stem-cell lines that were derived under more optimal conditions." Scientists will get unpolluted cell lines, which are free of animal products. Those lines offer "greater genetic diversity and are better suited for, still distant remedies, in humans." Stem cell lines created until May 1, 2007, can now be used for research in Germany. It is only allowed to import stem cell lines, their creation is forbidden.

On March 9, 2009 President Obama lifted President George W. Bush administration's eight-year ban on federal funding of human embryonic stem cell research. Previously to 2009 federal research money was limited to stem cell lines created before Aug. 9. 2001. Private supporters were not bound to the previous federal ban. Their contribution developed human embryonic stem cell lines. Dr. Georg Daley of the Harvard Stem Cell Institute and Children's Hospital of Boston petitioned to lift the strict limits. The leading researcher still considers embryonic stem cells as the most flexible and thus most promising form. He thinks that science and not politics should ultimately judge. "Science works best and patients are best served by having all the tools at our disposal," stated Daley.

However, some researchers see no problem in working with aged stem cell lines. Shinya Yamanaka, to whom we owe the discovery of induced pluripotent stem cells, is only working with human embryonic stem cell lines created in 1998. Bodo Ekkehard Strauer is a prestigious adult stem cell pioneer expert. In 2001, he developed the worldwide first cardiac infarct treatment based on adult bone marrow stem cells. So far, his method has been used successfully in more than 1000 patients. For Dr. Strauer, promises of human embryonic stem cell therapies are nothing else than wishful thinking. The fact is that human embryonic cells generate malignant tumors. The hidden nature of stem cells will finally reveal the false promises (Magnis 2008). Embryonic cells are growing as fast as tumors. A tumor treatment combined with human embryonic stem cells and tumor-controlling medications will most likely kill the embryonic stem cells. This is not at all the desired therapeutic way to go.

5 In Vitro Fertilization

During the 1950's, Robert Edwards a Manchester-born physiologist studied the development of embryos in mice. His research was focused on how hormones control critical ovarian function such as egg maturation and ovulation. Edwards could demonstrate that egg cells from rabbits could be fertilized outside the animal body. For twenty years he studied the life cycle of human eggs and tried to solve a series of problems in getting egg cells and sperm to unite and mature. Finally in 1969 Edwards and his colleague Dr. Patrick Steptoe, a gynecologist and pioneer of laparoscopic surgery, were able to fertilize oocytes outside the human body. Edwards recalled the moment: "I'll never forget the day I looked down the microscope and saw something funny in the cultures. I looked down the microscope," Edwards repeated, "and what I saw was a human blastocyst gazing up at me. I thought: "We've done it." He realized the potential of in vitro fertilization, said Christer Hög, Professor of Cell Biology at the Karolinska Institute in Stockholm. After another decade Edwards overcame "step by step" the technical hurdle and was finally able to show that oocytes could undergo in vitro fertilization and in vitro maturation to rise to early-stage embryos and blastocysts. In 2010 Edwards was awarded with the Nobel Prize (Alok 2010). Before Edwards could try to have his first test tube baby delivered the British Medical Research Council, a government-funding agency, rejected the researcher's request to try to make the technique safe for humans. Edwards and Steptoe even thought of emigrating to the United States. Private funds eventually helped them. For many years now a tough debate has raged over in vitro fertilization. Ethicists told that Dr. Edward is misleading the infertile. Like any medical procedure there are huge risks. Abnormal babies would threaten the health of the participants. The whole method would depend on women agreeing into the procedure. They would further acquire human embryos for research. Edwards was aware of the ethical issues. While defending his work in public, his labor was burdened with daunting problems. It looked so easy to mix eggs and sperm in a Petri dish and surrender the rest to the nature. But in reality, Edwards spend two years trying to get eggs to mature outside the body. It was reported that eggs would mature in twelve hours, eventually he learned that at least twenty-five hours were required. In this light, the criticism seemed not so wrongful that the normal order was being subverted by unnatural intervention in the mysterious process of creating a human being. From the very beginning, Dr. Edwards was in need of a reliable supply of human eggs. He approached Dr. Steptoe because of his expertise in retrieving unfertilized eggs from the ovary through minute incision in the patient's skin.

In 1972, they began transferring fertilized eggs to the womb. They hoped the rate of implantation would be as high as with farm animals. The scientists continued with the application of hormones given to the mother to induce ovulation. It turned out that the hormones interfered with the growth of the embryo. To improve the conditions the doctors injected mothers with extra hormones, but only achieved that the children were aborted. The result after forty embryo transfers was an ectopic pregnancy, which Dr. Steptoe aborted. The mother of that aborted child was Louise Brown. In her second pregnancy Louise Brown gave birth to her daughter Louise Joy Brown (Wade 2010).

On July 25, 2008 the first test tube baby Louise Joy Brown

Tube Baby

celebrated her thirtieth birthday. Her journey toward personhood began when her mother wanted to find out why she couldn't conceive and eventually made her way to Drs. Patrick Steptoe and Robert Edwards. The fertility specialist peered through a microscope and carefully created her in a glass petri dish. Her conception was made possible outside the female body.

The world's first successful "test-tube" baby born via cesarean section in Great Britain heralded a triumph in medicine and science.

In the United States, the first in vitro fertilization took place in 1981. A girl named Elizabeth Carr from Norfolk, Virginia was born on December 28th. Some wondered if it was more luck than science. The most important question was whether the babies are healthy. Was the egg harmed in being outside of the womb, even for just a couple of days? Will the baby have medical problems? Millions of couples try to conceive a child. Is it justified that parents and doctors have the right to play with nature and thus bring a child into the world? When does life begin? If it is coming into existence at the moment of conception, are doctors killing potential humans when they discard fertilized eggs? It is necessary to harvest and fertilize several eggs to assure positive results. These are only some examples of key ethical questions related to in vitro fertilization during that time. In Germany 100,000 tube babies came into the world before 2002. In 2007, the yearly number of in vitro fertilizations was estimated to be more than 3 million.

Involuntary childless women are not necessarily medically infertile. New reproductive technologies lifted the pressure of marginalizing females. Women drew hope in being able to conceive their own baby. Ovarian screening studies are the first step to consider if a therapeutic intervention for infertility or sterility will have a successful outcome. Tests quantify the ovarian reserve follicles. A diminished ovarian reserve is associated with a low ovarian response. Premature ovarian failure, defined as loss of ovarian function, can occur before the age of forty (Muasher et al. 1988). In vitro fertilization (IVF) does not offer much help to people who confront this cause of infertility in which, due to age, no viable eggs exist anymore. Egg cell donation has allowed some women to become pregnant using donated eggs.

In Germany, in vitro fertilization is only permitted to married couples and not allowed for woman over forty. Egg cell donation is forbidden in this country, while everywhere else pregnancy rates dramatically improve with the use of donor eggs and embryos. Twelve percent of IVF relied on egg donations in 2005. Since 1996, the U.S. Center for Disease Control and Prevention noticed a four percent increase of women considering donating eggs. Not every woman can donate eggs and the policy is to increase and ensure a safe process for donor and recipient. Some programs prefer to use donors who have already given birth or successfully donated eggs, as it is believed that they are more likely to be fertile. It is suggested to be easier for them to deal with the issue of having genetic offspring born to someone else. The "donors" are in their twenties and thirties. The well-selected donors undergo a competitive psychological screening and fill out lengthy profiles. Twenty-one pages inquire about religion, hobbies or what occupation the volunteers have. Donating eggs requires a certain intelligence quotient. After a series of injections and the potentially risky procedure of egg extraction, the donors leave behind a dozen of their eggs. In Wisconsin the usually anonymous donors are paid a fee of $5,000. Whereas the amount can be several times higher elsewhere. Each fertility clinic has its own pool of egg donors. They also use agencies that recruit egg donors, such as "The Stork Society" with offices in the two American cities, San Diego, CA and Madison, WI. Suitable egg donors and the *bona fide* receivers are happy. Despite that, they do not know each other. Then again this might be the better solution. Which donor wants to run into the woman at the grocery store, to whom she sold her eggs? Donors do not know how their eggs are going to be used. Several consider it like giving blood. Somebody else needs it and the donor has it available. For decades, donated sperm were also widely used. Donors earn about $100 to $200 (Wahlberg a 2008).

An egg donor undergoes the same procedure as a prospective mother with in vitro fertilization treatment. All-inclusive evaluation and treatment include blood tests, ultrasounds, and injections to halt ovulation, other shots to grow multiple eggs, more injections to trigger ovulation when the time is right, and the egg extraction. The anesthesia for the egg extraction can carry the risk of complications. To stimulate the development of multiple ovarian follicles daily subcutaneous hormone injection is necessary over a period of seven to ten days. Mature oocytes are retrieved under ultrasound guidance by the insertion of a needle. Besides anesthesia complications, there are other possible side effects like an ovarian hyper-stimulation syndrome. That means, enlarged ovaries produce excess fluid in the abdomen and lungs. The severe health conditions can require hospitalization. Another adverse effect can be that the donor herself might become sterile or even die. The clinical follow-up visits are time consuming. Donors accept the inconvenient and potentially fatal health risks associated with the procedure at least partly for financial gain. The eggs will be mixed with the sperm from the intended father. If embryos result they will be grown in a lab dish before one or more are transferred into the uterus of the recipient. If a child is delivered, it is genetically related to the donor but the birth mother is the legal mother of that child. Occasionally advertisements are on behalf of a specific couple, which will offer a large amount of money to the right egg donor. They seek comprehensive donors with special qualities such as above-average height, athletic or musical abilities. Intended parents research many egg donor agencies and Websites before finding the right match, their most innovative and talented super-donor. Donors are not accepted who are at risk from exposure to HIV or other infections says a brochure called "guidebook for egg donors". The New York State task force on life and the law advisory group on assisted reproductive technology prepared the guidebook in 1998. Infertility specialists, consumers, ethicists and representatives of the American College of Obstetricians and Gynecologists found that frequently egg donors are not adequately informed about the process. In most cases the medical bills for procedures involved in the donation will be paid. Though, if long-lasting medical complications develop, the insurance of the donor will be billed. Usually no financial compensation is given to a woman who donates eggs to a relative or friend. Recently, monetary incentive for human research subjects and donors is common practice to advance new scientific discoveries. Whereas, according to the Internal Revenue Service, taxes have to be paid on any money received for donating eggs. Beside all of that, the issue of compensation for human eggs has become an extremely controversial topic. In nations such as Canada and the United Kingdom, payments are banned. While indeed, donating eggs is time consuming. The ethics committee of the American Society for Reproductive Medicine cites that egg donors spend fifty-six hours in the medical settings. The estimate includes: interviews, counseling, and medical procedures (Steinbock 2004).

To increase the success rate of delivering one (or sometimes accidentally more) desired child, the sterility therapists administer anti-estrogen and gonadotropins. Some researchers see a possible association between in vitro fertilization and cancer development. They argue that the medications used for ovarian hyper-stimulation boost circulating gonadotropins, which are responsible for ovary cancer. In 1990, a study showed newly originated borderline ovary tumors and ovary cancer attributed to exactly that treatment (Whittemore et al. 1992, Rossing et al. 1994). In this regard, it has to be considered that females with sterility and infertility treatments undergo health checks more frequently which increases the possibility of detecting pathologic abnormalities (Dor et al. 2002, Lerner Geva et al. 2003).

There are questionable boundaries between what are ethically acceptable methods and what breeches human rights. People shop for donor sperm and eggs like for consumer goods or the way one shops online for matchmaking sites. As soon as the US group "Single Mothers by Choice" discovered, to their shock, that several of them used the same sperm donor and therefore their children were half-siblings, they started to set up a voluntary donor register. Unfortunately, some donors are very popular. To take them out of the catalogue will increase the interest of others and mothers advertise to buy any unused vials of sperm from. A study conducted by the Institute for American Values, called "My Daddy's Name is Donor" (see www.familyscholars.org) showed that adults aged eighteen to forty-five conceived through sperm donation are hurt, confused, depressed and feel less trust in their parents. They are more disturbed when money was involved in their conception. Children, who will never have the chance to be raised by their biological parents, commonly use the phrase: "My sperm donor is half of who I am." An important topic is to confirm a child's right even to have knowledge of the biological parent. It is needless to mention that many fear being attracted to or having relations with someone to whom they are unknowingly related. There are no laws declaring anonymous donation to be illegal. Thus women will continue to travel to Spain or Romania to secure human eggs and to return pregnant to their homeland after cheap fertility treatment (O'Brien 2010).

5.1 Redefining Nature

Many attempts were undertaken until in vitro fertilization became successful. Sometimes a dozen embryos were prepared in order to obtain a pregnancy. In 2005 the success rate increased by thirty-four percent, which is twenty-eight percent higher compared to 1996. Younger women achieved better results with infertility treatment. Depending upon the quality of sperm the decision is made to accept the classic IVF or to apply the intracytoplasmic-sperm injection (ICSI) established in 1992.

With the more common classic IVF method, sperm and egg are incubated in culture media at a ratio of about 100,000:1. A time frame of eighteen hours makes sure that fertilization will take place on its own. In situations where the sperm count is low, a single sperm is going to be injected directly into the egg. The new ICSI might be even similar to smart breeding in plants. ICSI is a very effective method to acquire fertilization of eggs. Embryologists select individual sperms in a tiny specially designed hollow ICSI needle. They are guided by specialized micromanipulation equipment and inverted microscopes. The needle is carefully advanced through the outer shell of the egg and injected into the inner part of the cytoplasm. It has a success rate of seventy to eighty-five percent. In spite of everything, several eggs are consumed by in vitro fertilization and ICSI. One obvious difference between both methods is that malformation of the embryo does occur at a lower rate with ICSI. ICSI implementation is not the standard of care and is used only for severe male infertility. Nevertheless, more often IVF clinics are asked to perform ICSI. The treatment itself will add $1000 to the age-dependent normal costs. A thirty-five year-old female has to pay $22,000. A thirty six year-old female has to pay $26,000. The procedure for older woman gets more expensive. Women over the age of thirty-nine are even required to accept egg donations and younger candidates are encouraged to donate surplus egg cells (oocytes). If this is the case donors will definitely be able to benefit from doing so, in view of the fact that a discount will be offered to them.

It is becoming more common for IVF clinics to provide pre-implantation genetic diagnosis (PGD), also known as embryo screening. First used in 1989, it refers to procedures performed on embryos prior to implantation. The vast majority of infertile couples think PGD is an alternative to prenatal diagnosis. They argue that prenatal diagnosis might carry the risk of miscarriage. Otherwise, if through prenatal diagnosis a problem is detected in the fetus, the woman is faced with the difficult decision of giving birth to a baby with a genetic disease, like Down syndrome. Who wonders about the attitude of parents, who underwent the "exertion of IVF", to be reassured that their most desired child is all right? They demand the quality control of PGD more appropriate to the production line than the generation of human offspring.

The evaluation selects an embryo free of certain diseases. This technology can be conveniently combined with the creation of embryos. The clinic advises not to look for specific diseases. Instead they provide a technique to create embryos free of genetic abnormalities. In our times, many markers are available. The embryologist searches for genes causing cancer later on: Alzheimer, Asthma, Multiple Sclerosis, Arthritis,

Cystic Fibrosis or Sickle Cell Disease and so on. Only embryos free of the genes for the disorders or clean of anatomical or physiological abnormalities are considered for implantation in the prospective mother. Of course, hereditary family diseases stay in the screening focus as well. As a result, a twenty-seven year-old English woman gave birth to the first breast cancer gene free baby. The doctors at University College Hospital in London had created eight embryos through in vitro fertilization and screened them for the variant of BRCA1 breast cancer gene. Two embryos were chosen using PGD. Britain's first cancer gene-free designer baby was born on January 9, 2009. Women who carry this genetic variation have an eighty percent chance of developing breast cancer and a sixty percent chance of suffering ovarian cancer during their lifetime. In the described case the family of the father had carried the breast cancer disease for three generations. The father might have been spreading a heightened risk of acquiring this disease on his daughters. It is hoped that his little girl will not face the specter of developing this genetic form of cancer in her adult life. Although the daughter who developed from that embryo most likely will not carry the gene, it does not mean she would definitely not contract the disease. Many diseases are triggered by a complex of interactions between multiple genes and environmental factors. Nevertheless, the Human Fertilization and Embryology Authority approved the procedure of specifically selecting embryos that did not carry the gene (Devlin 2009). Still, the procedure has proven to be controversial, because otherwise healthy embryos that carry only the defective gene are discarded. People have raised questions about the ethical acceptability of using genetic characteristics as selection criteria. Doctors justify the screening of test-tube embryos carrying diseased genes. They consider it as a specific treatment. This is similar to the technical implication of PGD undertaken to test tissue-typing, when parents of a sick child are given permission to have another healthy baby with matching tissue such as bone marrow. Professional authority argues that in very rare circumstances and under strict control a child could be born following PGD. His suitable cells would match that of his sick sibling and could be used in a transplant operation. Hereupon the Human Fertilization and Embryology Authority did extend the ethical guidelines on the use of PGD and doctors were allowed to choose embryos because of their genetics. It seemed irrelevant for reproductive experts that this act inflamed the debate or even lead to the creation of "designer babies" (Connor 2001).

However, pre-implantation genetic diagnosis is the subject of a heated debate and considered by many as morally inappropriate. Nobody should have the right to kill healthy surplus embryos in order to design a cancer-free female baby, or a suitable tissue donor. Especially under the circumstances in November, 2008 when the data analyzed by the National Birth Defects Prevention Study showed that assisted reproductive technology has brought forth major structural birth defects. During a period of October, 1997 to December, 2003, infants were screened for major birth defects whose mothers used IVF or ICSI. The results were compared to control mothers who had unassisted conception. Artificial conception was associated with septal heart defects, cleft lip with or without cleft palate, esophageal atresia, anorectal atresia. The findings suggest that some birth defects occur more often among infants conceived in vitro (Reefhuis et al. 2008). For pre-implantation genetic diagnosis, embryos have to be created in vitro. This technique itself might provoke birth defects. In this context it seems incomprehensible to destroy three day old embryos who carry a genetic defect. Beyond the harm of embryo destruction accompanied with PGD, scientists see the potential to damage the embryo by isolating one of his cells and fear that an implanted seven-cell embryo will not come to term. Ethicists raise the issue of destroying an identical twin to the implanted healthy and suitable embryo. To go in *medias res*, pre-implantation genetic diagnosis is accomplished on day three in the eight-cell stage, where every cell is still omnipotent. One of the eight cells is separated for analysis of its DNA to determine whether or not the embryo is likely to develop a genetic disease.

The technique of removing individual cells from an embryo is a commonly used procedure in animal breeding. With artificial embryo splitting, the mostly bovine embryo in the eight-cell stage is separated into single cells. The embryo was either removed by non-surgical flushing of the uterus or used after IVF. The cells are then placed into petri dishes that contain the correct nutrients and hormones for growth. Each of them divides to a new embryo. These embryos will be transferred into the uterus of surrogate mother cows to deliver later on as identical twins.

In spite of everything, assisted artificial conception depends in some cases on donated egg cells. In this context, it is imperative that the information from the egg donor regarding inherited diseases is correct. It is argued that PGD can be especially useful for IVF patients who know they carry genetic disorders. However, ethicists have warned of a slippery slope in which embryos might be picked because of other traits. Medical interventions associated with human reproduction give rise to strong views about the social acceptability. The word "selection" has a bad reputation in European history. In Germany pre-implantation genetic diagnosis was by law forbidden due to the Embryo Protection Act, which came into force in 1991 until July 6, 2010. Restrictive bio-politics about ethical implications of the Human Genome have been vigorously debated. In Germany PGD, egg donation and surrogacy are illegal as well. Additionally the system of funding infertility treatment by health insurance has changed recently in this country. Couples have to pay more for IVF. As a result a certain amount of "PGD" tourism occurs. British couples who want a girl instead of a boy must go abroad. The most common choice is to fly to Jordan, where sex selection service is done in affordable price of 2000 English pounds, including accommodation costs and flights.

Researchers think that Germans have a considerable knowledge deficiency regarding the possibility of modern reproductive medicine when they consider PGD screening for sex selection and other, non-disease related reasons. "People know about the existence of PGD and fertility treatments, but do not necessarily understand their applications and limitations and overestimate the diagnostic possibilities of PGD," said Dr. Ada Borkenhagen, a psychologist and researcher at Charité Berlin, in a study done together with colleagues at the Berlin Fertility Center and the Universities of Leipzig and Berlin. The observation further revealed that twenty percent of the general population think PGD could be used to indicate future characteristics such as height, eye and hair color (Mason 2004).

In 2010 a German gynecologist who prefers to do his job in the test tube decided to implant only healthy embryos and thus performed PGD to detect genetic diseases. He was motivated because the current legislation on pre-implantation genetic diagnosis in his homeland is, as he believes, out of step with the attitudes of his fellow Germans and should be changed. The physician wanted to make sure to insert only the healthy ones of the three created embryos. At present creation of only three embryos is permitted and all of them have to be implanted to give them a chance to develop. Afterward he sued himself. On July 6, 2010 a German court decided that the reproductive physician is not guilty. This decision has quite an impact and might initiate German legislators to consider changing the law. It seems that by the very nature of this quite challenging topic the discussion will become compassionate and emotional. Dalai Lama wrote: "True compassion is not just an emotional response but a firm commitment founded on reason. And we all know that when we are ruled by emotions all truth, reasoning and logic is brushed away."

No doubt, ethicists are well aware of the argument that pre-implantation diagnosis could be justified so as to be in the interest of embryos, who would otherwise have been born with a serious genetic disorder. Dr. Helene Watt, a senior researcher at Anscombe Bioethics Center in Oxford, England, emphasizes that such a statement unmistakably claims that the existence of a disability makes a person bad or worthless. "This indicates quite an extraordinary judgment of a fellow human being." In Dr. Watt's opinion it is rightly said that "good parents are parents who accept their children unconditionally, for better or for worse". Pre-implantation diagnosis interferes with that "statement" and rejects children in case they are "worse". Watts reasoned that parents might reject those of their born offspring who fall below a certain standard. She determined that society in general will be affected by pre-implantation selection. "We are almost inclined to think that older disabled people live in a society which sees their condition as a fate worse than death or at least for selecting out before a certain stage of development," reported Dr. Helene. "We live in a time when it is undeniable that fellow humans with certain types of diseases already receive the message that life with such conditions is intolerable for the individual and/or for the family," concluded the expert (Watt 1997). The regulation of reproductive medicine technologies differs among European countries. Great Britain allows this procedure only under the supervision of the state institution: Human Fertilization Embryology Authority (HEFA). Synonymic the UK regulator for fertility treatment and embryo research is licensing some clinics to conduct the PGD tests. The technique still recalls the memory of the Third Reich and its hereditary discrimination. Modern medicine makes it possible that the embryologist decides who is going to live. The brothers and sisters of the designed and desired child are ending up as bio-waste. Economic pressure is an excuse that nobody would wish to have a handicapped child produced by IVF. In case PGD fails, since other factors are involved in the development of a healthy child (like epigenetic modifications of the DNA), and result in the birth of a disabled child, insurances need to reimburse the parents. Does that indicate that a disability is judged as a reduced and unworthy quality of life? The ethical value of life itself seems diminished. The phrase "Unwertes Leben" (unworthy life) had already been used by the Nazis. Germans are still stigmatized by the Nazis. "It is incomprehensible for Germany that it appears to happen again, when embryologists have to conduct selection during prenatal diagnosis," recently remarked a well-known German cardiologist. Contrary to that kind of selection is the insight of E. Frankl who stated: "Every new human being who comes into this world is an absolute novelty, since the spiritual existence cannot be inherited or passed on by the parents to the child. In procreation, we can only pass on the building blocks. It is impossible to inherit the spiritual being."

"Some people are convinced that an embryo at the moment of conception and beyond is not human. He is not worthy to live, because he cannot reason and has no feelings. Consequently, the step of declaring an impeded person as not privileged to live is not far away," argued the adult stem cell pioneer Dr. Bodo-Eckehard Strauer, chief of the Cardiology Medical Center at Düsseldorf University/Germany (Strauer 2008). "Do we really want a genetically cleaned up future?" questioned

Baron Dr. Tom Shakespeare. He is a geneticist and sociologist with an appointment at the University of Sunderland, United Kingdom. He himself has achondroplasia (shortening of limbs). He is a campaigner for disability rights and bio-ethics expert. In former times, he fought against   to abort handicapped children. "In our days it is considerable when a mother brings her child with special needs to term in the face of discrimination. Prenatal diagnosis has the potential to be eugenic. During the Nazi time it was the dictate of the state. In our days it is the freely willed decision of the parents that leads to the killing of an unborn disabled child", says Shakespeare. He continued: "People who are against the birth of a handicapped child believe it has no right to live because in their opinion it has no quality of life and thus its fate should be spared. It is also considered an avoidable burden for society. In general, humankind is very troubled to see meaning in suffering itself. People are scared to endure psychological or physiological difficulties in raising a disabled child. The pressure of society to live in such conditions is viewed as extremely challenging" (Schneider 2002). The renowned fertility expert, Dr. Anthony Caruso quit his job after realizing his participation in an increasing objectification of children. Dr. Anthony Caruso was initially motivated to dedicate his work as a reproductive endocrinologist. He wanted to bring happiness to an infertile married couple. Instead he realized that: "the procedure is at odds with the self-sacrificial ideal of marriage. It's...the idea that you can have whatever you want, wherever you want, whenever you want," Caruso told EWTN News on June 9, 2011. His colleagues ridiculed him when he revealed that he could no longer work as part of the "increasing objectification of children". However, Caruso deeply regrets that he had wronged other people and had lead so many couples down a road that was wrong. Horrified, he quit his job (Gilbert 2011).

Whereas, the ethic professor John Harris, Manchester University, Great Britain is convinced that the basic law of a democracy should be applied in the exercise of the free choice regarding everything that concerns reproduction. He has the opinion that every biotechnology, that is available and desirable, like a child's features according to the preferences of the parents, should be allowed. The parent's decision should be paramount (Schneider 2002).

In contrast the broad public were shocked when Nadya Suleman gave birth to octuplets in Los Angeles in January 2009. A Beverly Hill fertility clinic faced a violation of the standard care for implanting so many embryos. Dr. Michael Kamrava, the IVF practitioner lost his medical license on July 1, 2011 for respecting the request of the mother to transfer all of her remaining embryos into her womb. By doing so, the IVF practitioner violated and was not in compliance with the guidelines.

National guidelines put the norm at two or three embryos in order to lessen the health risks to the mother and the chances of multiple births. The single mother stated that she had six embryos left over from her five previous pregnancies. This time she wanted all her remaining embryos provided by in vitro fertilization to be implanted. She did not want the doctors to selectively reduce the number of embryos. The mother always had the desire to give so many siblings to her other six children. She further refused the option of a selective abortion, when told she was expecting multiple babies. The subsequent delivery, while being only the second time in history that octuplets have been born and survived, caused worldwide attention. For the mother it would have been more painful to kill her preborn babies than to bring them to term. Dr. James Grifo, professor of obstetrics and gynecology at the New York University School of Medicine stated: "I don't think it is our job to tell them how many babies they are allowed to have. I am not a police officer for reproduction in the United States. My role is to educate patients."

Consequently, the mother did follow the technical advancements of IVF. Some commentators immediately demanded revocation of the license of the doctor who performed the IVF. The press urged doctors to abort the children or otherwise pay child support until all eight innocent infants reach the legal age.

Suddenly IVF is viewed as an irresponsible act, as it is feared that taxpayer's money is going to support a single mother of fourteen kids. Can a physician "out of the blue" deny a mother the so-called "reproductive rights"? Is it forbidden for poor people who do not have a job to have children? However, the fact that taxpayer's money is already used to cover abortions and human embryonic stem cell research does not cause public outrage!

Is the so-called "pro-choice" attitude a one-way street by only supporting the decision against children? The capability of having a designed child however, can be also considered a justification to abort an unwanted child. There seems to be a dilemma in the focus on either the well-being of the children and the increasing possibility of parents to have a certain right to design their children. Medical ethicists appear concerned about the well-being of octuplets, but not about embryos that are destined to be thrown away because their eyes are brown instead blue.

In 2009, a Los Angeles fertility clinic offered for the first time a new service to design babies according to the parents' exact requirements. The fertility institute gives parents the chances to select. They can request a son with brown eyes, black hair and a dark complexion, or a pale, blonde, green-eyed daughter. Other fertility specialists are outraged. The clinic seeks to capitalize on advances in embryo cell analysis aimed at identifying diseases or defects with pre-implantation genetic diagnosis in the unborn. Medical technology can hence produce children free of debilitating genetic conditions. This act is justified by the progress of science. The recently dramatic advances in the ability to analyze cells makes it possible. Medical genetics are able to get enough DNA from a cell to identify thousands of characteristics of the embryo. The Los Angeles in vitro fertilization clinic is already the world's largest provider of prenatal gender choice. Marcy Darnovsky, director of the Center for Genetics and Society is very much concerned because we are going to create a society with new sorts of discrimination. "Today it is the eye and hair color. Tomorrow it is the height and intelligence" (Sherwell 2009). One might argue that just because something can be done, doesn't mean it should be done. Also a certain irrational attitude is increasingly observed in the sense that modern biotechnology is in some kind of control. Accordingly, when an individual is in charge, he/she has a certain right to receive perfect service in the form of a total healthy, super intelligent child.

5.2 Transgenic Pets

Genetic engineering is a high-tech method to breed desirable traits in offspring, or to deliver future generations from undesirable diseases. For example the insertion of a genetically engineered and more efficient foreign gene into a specific gene locus, that can improve breeding and farming results. Genetically modified goats can produce milk containing spider silk. One of the wonders of nature is spider silk. A material tougher than Kevlar, a plastic fiber used to make bullet-proof vests. Bioengineered goats can produce milk that contains the protein that spiders secrete to make their silk. The Air Force Office of Scientific Research is cooperating with a team from the University of Wyoming to investigate spider silk proteins. The team will obtain useful quantities of a new kind of silk. Nevertheless, "To make a five-pound bulletproof vest, a producer would use 600 gallons of goat milk containing the silk protein, The milk production from 200 goats in one day would be used for just one vest", explains Dr. Randy Lewis, team leader of the experiment (Callier 2008).

Veterinary medicine research usually is the trailblazer for human medical science. Bio-Genica, a genetic engineering factory, claims to be able to produce pets developed through genetic engineering. The so-called Genpets™ are attributed to recent veterinary medical innovations. Genpets™ are allegedly genetically modified animals optimized for modern household and living. It is understood that the company delivers prepackaged, bioengineered pets. Generally speaking they are ostensibly living, breathing genetic animals. The completely unfamiliar process of zygote microinjection (ZMI) brought the animals to life. Researchers claim a revolutionary breakthrough in the method to combine DNA or certain proteins. Finally they insert them into the species of the client's desire. In 1997 the method was purportedly used to accomplish the task of generating the bioluminescent jellyfish-mice. Once they have taken that first step the applications went straight upwards, from glowing pigs, rabbits, fish and monkeys. Before you know it, they will insert human DNA into rabbits or chimpanzees or spider DNA into sheep, to commercialize Genpets™. The company stated that National Geographic needed many additional reprints to satisfy the demand of readers. Right now the delivery of the Genpet ™ is still free which causes many peoples to hurry to get their Genpet ™ in its unique plastic package (Bio-Genica 2008). Of course this report is a very funny joke. Even so one might wonder when that technique will be applied to humans. Nevertheless, the method described is thought provoking. "Genetic engineering is very controversial, it raises issues of sex selection," a fact sheet of the American Society for Reproductive Medicine quoted. Genpets™ show that the dream is still alive to redefine nature in a convenient fashion, which fits totally into our hectic world.

Fertility clinics already offer various reproductive services. The desires of the females are diverse. Fertility medicine is recording an increasing use of egg freezing in liquid nitrogen to be stored for future use. This procedure is equitable to the freezing of embryos or sperm. Egg freezing service is tailored to two groups of women: cancer patients, whose ovaries can be destroyed by chemotherapy and radiation, and women in their twenties or thirties who want to defer pregnancy. The service, which requires an annual storage fee of $440, can free women from some of the biological constraints of fertility. It allows them to pursue motherhood when their relationships or career is optimal for a child. In the nitrogen tank are also surplus embryos. Their storage costs are high. Eventually they end up as waste. Employees of IVF clinics who have this assignment consider it a heavy burden, for which nobody likes to volunteer (Pacholczyk 2006). Neither are genetic parents in favor of their frozen embryo-children being adopted and carried to full term by other women. Adopting a born child is already difficult enough. Imagine a frozen embryo adoption agency, perhaps with the name "nitrogen-kids". Nowadays, embryos are in some cases heartily welcomed. But their existence is also considered undesirable because of his or her genetic traits and we easily assess them as unwanted bio-waste.

Immanuel Kant a German philosopher discussed in his book "Critique of Aesthetic Judgment" the four possible reflective judgments: the agreeable, the beautiful, the sublime, and the good. The agreeable is based on our senses. The good judgment depends on ethical decisions that confirm the moral law. It is a purely objective judgment because things are either moral or not. Subjective judgments are the beautiful and the sublime. Subjective decisions are made in the belief that other people ought to agree in these judgments. Kant is speaking from a "sensus communis" which he calls the basis or the community of taste that will combine people to recognize certain things. The establishment of the free will and the faculty of mind and reasoning are dependent on our judgment. The ability to judge is thus given by our intelligence. "Our organized bodies necessarily seem to us as though they were constructed according to a conception of purposes which preceded them," said Kant, and concluded that: "our bodies were made by a purposeful intelligence like ours" (Pluhar 1987). Can we assume that intelligence was given to us when we were created? Kant's basic question was: "How can we know truth"? He found the answer that we know truth by the intellect, not the senses. "I have to make decisions after my free will and according to my intellect and cannot be guided by emotions which destroy the truth," said Peter Kreeft, a well-known philosopher in our times (Kreeft 1990). Everybody loves to make judgments in regard to human embryonic stem cell research, which are more or less based on sensations, or emotions. In case a fire breaks out in a fertility clinic the fireman will more likely save a nurse than frozen embryos. Nobody would even consider risking his life for embryos, especially in circumstances which lead the rescuer to make his decisions under pressure. He is forced to make life and death decisions whether to save frozen embryos or a nurse. In such a situation a fireman reacts to what is familiar to him. In a moment of crisis he has no time to engage in logic or rational decisions. Furthermore, embryos in a frozen tank are unfamiliar to us. Saving the nurse makes sense in a delicate situation, because he justifies his act with the outcome of saving a life. The future ability of embryos to live is uncertain. It is seldom that embryos from the frozen tank will end up being implanted into their mother's womb. Rather, they might be handed over to researchers for embryo-destructive experiments. Nevertheless, an unrealistic scenario does not justify our acts. It is not testimonial evidence that an embryo has no intrinsic value or dignity because the rescuer had an excuse to save the nurse instead of them (Pacholczyk 2008).

5.3 Womb for rent

Anand, a city of 130,000 in the western Indian state of Gujarat, churns out babies for the wealthy countries. The "rent-a-womb" industry is a fast-growing, multimillion-dollar commercial surrogacy business. India does not have guidelines yet. Anything can be done. Carrying a child for rich United States or Japanese couples is an effective sales pitch. Each day women show up at the many Indian fertility clinics, hoping to be hired as surrogates. The India Council of Medical Research continues to urge the government to promote new legislation. Laws could prevent gay and lesbians from hiring surrogates, since India has a legal ban on gay marriage. With the force of law, couples would be required to buy a nine-month health insurance, besides a temporary life insurance for surrogate mothers. The ideal surrogates must be between twenty-five and forty. They are not allowed to have more than two children of their own and have to accept "foreign-embryos". Husbands must agree and the participant has to be free of tuberculosis, diabetes, high blood pressure and HIV/AIDS. All that a couple needs is to provide their egg and sperm and $ 7,000. Panash, a thirty year-old mother, chats about her experiences. Last year she carried a child from a Japanese couple. It was conceived with the husband's sperm and an anonymous donor's egg. The couple divorced and the ex-wife decided not to have the child. To obtain a passport for the child, it is necessary that the mother is present. Surrogacy is uncharted territory. Legal and ethical complications often hinder "genetic" parents in bringing their new children home. A Canadian couple went to the Canadian High Commission in New Deli for travel documents for their twins. The officials ordered a DNA test which failed and showed that the babies were not related to the Canadian couple. They were the "in vitro product" from a different unknown couple and are now sentenced to spend their childhood in one of India's thousands of orphanages. Travel documents, citizenship, the possibility of treating surrogate cases as adoptions, whereby the child first gets an Indian birth certificate, are perplexing problems for embassies. They don't know how to handle babies with two sets of parents and two nationalities. Surrogate child-bearing lets the Department of Foreign Affairs worry about potential cases in which the Indian women claim that they were exploited by fertility doctors and western couples (Westhead 2010).

Did anyone think of this sort of scenario when in vitro fertilization produced Louise Brown in 1978? At that time doctors were convinced to help the infertile to have children. Nobody imagined using IVF just to get a boy, or a trait in a child that parents preferred to be passed on. The method is not about treating infertility; indeed it never has been, because the parents remain unable to conceive naturally, before and after IVF. It is about designing our descendants argued Dr. Arthur Caplan, the director of the Center of Bioethics at the University of Pennsylvania. Today people want a baby without a father, or in an advanced age, after they have had taken contraceptives for a long time. Couples have all sorts of reasons for wanting a baby. Did we raise objections when hearing that in India, parents in their seventies wanted a baby-boy. In 2008, Omkari Panwar, a seventy-year old woman from India gave birth via Cesarean section to twins, a boy and a girl after infertility treatment. The mother and babies underwent a super high-risk pregnancy. Panwar already has two daughters and five grandchildren. Even if modern medicine is able to pursue such a pregnancy and healthy babies are born, the children would not have a real chance of having at least one parent who is physically capable of raising them. And how can a treatment be identified as "infertility treatment" in a woman, two decades beyond the typical age of menopause. The Indian couple with the newborn twins wanted a boy. Such a statement claims that sons are better than daughters. The main question is, whether a seventy-year-old should be accepted as patient by any infertility program. Is the ability of medicine to make orphans a good enough reason for doctors to try IVF? Thirty years earlier Drs. Steptoe's and Edward's "mechanistic creation in the glass dish" was criticized for degrading human dignity. Some people argued that creating children by IVF is simply unnatural and would stigmatize the child. As reproductive technology evolves society get used to the idea of starting human embryonic life in a dish. The early pioneers struggled to get one egg from the ovary of the mother and so never had concerns about how many embryos to put into a uterus. But today it is common to make babies by numbers and to abort them prenatal if there are more than the couple's wishes to have (Caplan 2008).

5.4 Outside their power

Theodor Adorno, (1903-1969) a German-born international sociologist and philosopher and member of the so-called "Frankfurt School", (refers to a school of neo-Marxist interdisciplinary social theory) sees humankind in a constant fight against their own internal nature. According to Adorno a failure to discipline their own nature will be accomplished by misadventure and will prevent them from experiencing luck and happiness (Horkheimer and Adorno, 1988). In vitro fertilization was actually established to fulfill the wishes of some unlucky, sterile, unfortunate infertile parents. Unintentionally it opened up a completely new chapter. Bioethics, scientists, physicians, environmentalists, philosophers, theologians and politicians, almost everybody is very emotionally involved in discussing the consequences of in vitro fertilization. In vitro fertilization proves that the embryo is an independent being. It was created separately from the mother. In 1978, physicians were still concerned about whether the creation of an embryo outside the womb might harm its future development. Of course, the embryo cannot survive if it remains outside the uterus. Every living being depends on its specific environment and nourishment. People justify embryonic research with the opinion that the embryo cannot live by itself and thus has no value as an independent human being. However, nobody can blame embryos if they do not survive once they are taken out of their supporting environment. From the very beginning we are necessarily related to others. If this dependency is denied we can neither live nor develop. Humans are by nature social beings. Interpersonal communication originated in the companionship between male and female, beginning with the first parents "Adam and Eve". The very intelligent snake with the power to speak gave Eve the choice to determine whether or not she should disobey God. Well, disobeying God caused Adam and Eve trouble. They should have known that. But where do we as human beings find orientation in our time? I doubt that we will ever have the opportunity to argue with a reptile. You certainly can subscribe to a seminar about how to talk to your dog and you will be able to understand what he is saying to you.

Incarcerated parrot

Whatever you do, animals act by instinct and we have our intellect and free will which separates us from them. So what is the point?? Do major advances in medicine provide guiding principles? Humans have the ability of self-awareness and with that guidance through their own conscience. It is thought of as the ability to apply the law of human nature to a concrete situation in life. In the depths of his conscience, man detects a law which he does not impose upon himself. The voice of conscience speaks to him if necessary. Through it, he turns aside from blind choice and strives to be guided by the objective norms of morality. The norm of human activity should be in accord with and in harmony with the natural moral law (Fagothey, d 1958). And last, but not least, by doing so we experience happiness, to quote Adorno.

The big question remaining is whether humans can do everything according to their abilities even under the circumstances in which our actions are in disharmony with nature. Do we have the right to perform in vitro fertilization or to abort our children, just because we have the techniques and financial means available? There is a basic tension between concerns about the adverse consequences. Opposing opinions are common in modern scientific research. For some it is a deeply offensive act when science literally assumes the role of

God. Society fears that we may fail to develop important technologies if we apply too much restraint. Regardless of how one feels we must ask if we as humans, are allowed to become ruler over the source of human life? The challenge of new genetic technologies is to alter how we have babies and what baby we allow to come into existence. Many people consider whether ethical concerns surrounding this topic will be ever settled. Ethics originated in anticipation of the good life. For the Greeks ethics is related to social customs and categorized as a part of philosophy. It deals with rightness and wrongness of human conduct as known by natural reason.

Interesting in this regard is that Pope Paul VI published his encyclical letter "Humanae Vitae" more than forty years ago. Is it a prophetic statement since at that time neither abortion nor in vitro fertilization were legal, or even thought of? "Humanae Vitae" speaks about passing on human life. It was published on July 25, 1968, exactly ten years before Louise Joy Brown was born via cesarean section. Her birthday was also the birthday of the commercialization of in vitro fertilization. "Humanae Vitae" states, among other things, that humans are not rulers over the source of human life. "Man's service regarding co-creation is more seen in the cooperation and compliance with God's plan and his loving design. Humans never have completely unrestricted power over their body, especially not in the moment when they collaborate freely and responsibly with God the Creator. This act is far from being the effect of chance or the result of blind evolution of natural forces. It is the wise and provident institution of the Creator. In the procreative faculty, the human mind discerns biological laws that apply to the human person". Pope John XXIII said, "Human life must be holy to all. It demands from its very beginning the intervention of the Creator" (John XXIII 1961). In 1988 Pope John Paul II declared in "Christifideles Laici": "The inviolability of the person which is a reflection of the absolute inviolability of God, finds its primary and fundamental expression in the inviolability of human life. Above all, the common outcry, which is justly made on behalf of human rights for example, the right to health, to home, to work, to family, to culture is false and illusory if the right to life, the most basic and fundamental right and the condition for all other personal rights, is not defended with maximum determination."

Every human person has inalienable rights, the right to life, liberty and pursuit of happiness. These rights are endowed by the Creator and not by the State. These were rights recognized by the founding fathers of the United States of America to be given by the Creator to every human being and the American constitution does not limit these rights to only born humans. The State, or in case of abortion a mother does not endow this right to take her child's life, which she has not created. Stem cell researchers believe they are the creator of human life and thus eliminate the Creator. Therefore, they believe they have the right to destroy embryos for research purposes.

"Gaudium et Spes", the Pastoral Constitution of the Catholic Church in the Modern World, one of the chief accomplishments of the Second Vatican Council stated that without the Creator the creature will disappear. When God is forgotten, the creature itself grows unintelligible (Gaudium et Spes 1965). Are we not doing exactly this with stem cell research and thereby directly destroying the embryo and endangering our genetic future with experimentation on the genome?

Is it merely required that creating human life be devoted to the future of humanity and based on the moral and natural law imposed by the Creator? In vitro fertilization (IVF) and preimplantation genetic diagnosis (PGD) raise important concerns regarding its safety, effectiveness, costs and access. It is still unclear if and when it should be used and if one gene can become more superior to others. In the end this would be more a matter of choice than chance. Science seems to have few reservations or see limits based on the ethical, moral or societal acceptability. There is no policy about alternatives that could guide the future development of this technology. Genetically undesirable embryos are likely to be destroyed because nobody would freeze them indefinitely. "Reproductive technologies involve the creation and, frequently the destruction of human embryos. The method is therefore objectionable to people who believe that a unique human being is formed at the moment a sperm fertilizes an egg and deserves the protection of a born person. Thus PGD cannot be called therapeutic, because the testing does not treat the conditions it detects." PGD diagnosis is done with the sole purpose of telling parents which "patient" to discard. Those arguments surrounded the "advent" of IVF (Hudson et al. 2005). "There are some people who frankly think that PGD and IVF are unnatural or violating the ways of nature", noted the President's Council on Bioethics in its book, Beyond Therapy (Kaas 2003). The human mind has to discern biological laws in order to apply them for the benefit of humankind. Researchers want to analyze embryonic development in order to reproduce it. Science sees as one of its main tasks to prove truth. Whereas science can fail, truth will never be mistaken. The challenge for science is to learn how an embryo develops in the womb. Can Science imitate nature by disobeying the laws of nature?

Enzensberger, a German author, poet, translator, and editor is questioning the new biotechnologies of modern time. Science promises an extraordinary benefit for humankind by the use of human embryonic stem cells in research. Whereas, humility and skepticism in the face of science would be more appropriate for such undertaking (Enzensberger 2008). We need humility about our knowledge. As the knowledge of science constantly updates, corrects, and offers new insights, it contradicts prior scientific understanding. An atheistic scientist claims to only believe in the existence of things that can be scientifically proven. This attitude however assumes that things that are not provable by science do not exist. Yet in science itself, things that cannot be proven scientifically are claimed to exist because we know of their effects. For example, subatomic matter is claimed to exist. Even though it cannot be proven scientifically, only the effects of it can be observed. Why can we not make the same analogy with faith? And God whose presence we cannot prove scientifically, be recognized by the effects of his existence? We can recognize intelligent design in our "World". The size of the earth is perfect. A smaller or bigger atmosphere would be impossible. It is located in the right distance from the sun and is the only known planet with the right mixture of gases to sustain plants, animals and humans. The moon is perfectly placed and creates important ocean tides and is restrained from spilling. Evaporation takes the salty ocean waters and forms clouds which the wind blows away to disperse water for the living beings on earth. Water means that thousands of chemicals and nutrients can be carried throughout our bodies, and enables food, medicine, and minerals to be absorbed. The human brain is analyzing simultaneously a million messages a second, an unbelievable amount of information. You can go on and on and on. The astrophysicist Robert Jastrow (a self-described agnost) stated: "The seed of everything that has happened in the Universe was planted in that first instant; every star, every planet and every living creature in the Universe came into being as a result of events that were set in motion in the moment of the cosmic explosion (big bang). The Universe flashed into being, and we cannot find out what caused that to happen" (Jastrow 2002). We very well observe that the

Universe operates by uniform laws of nature that never change. They are totally reliable and orderly just as gravity remains 24/7. Life is built on the information of hereditary DNA. Can we as humans with our limited knowledge and expertise interfere in the process of life by the selection of our preferred embryos via gene-screening?

6 Designed children-the lesser evil

6.1 Frozen for breakthroughs in the future, no money no freezing

Life, the surface of the earth and the environment are constantly in a flux of new life and death. We have to adapt to the environment and change by a process of mutation when we want to survive and exist on this planet. "Although we may pass our genes to our offspring, we cannot pass on our memories or our conscious selves - those intangible qualities which make us individual entities - thus it is not easy to accept death knowing that we will cease to be," ponders the Physiology University Professor Dr. Muldrew from Canada in his writings about: "the study of living things at low temperatures," in 1999. He continues that if we were to be immortal, the population would soon rise to a level which could not be supported by the environment (Muldrew 1999). But are immortal humans required to reproduce themselves if this act is endangering their existence? We would not need the success of any advantageous mutations that makes our offspring slightly different from us. Maybe we do not need offspring at all and thus there is no need that two organisms combine their genes. But is it worth living without having children and missing out on sharing our love. Saint Valentine's day would be cancelled; we traded it in just for the comfort of immortality.

In 1964 Robert Ettinger published a book called: <u>The Prospect of Immortality</u>. He suggested that newly deceased could be frozen and stored in liquid nitrogen and, by implication, Robert promoted the concept of cryopreservation too. The word cryogenics comes from the Greek words "kryos" cold or freezing and "genes" what means born or produced. Robert Ettinger is a physics teacher, science fiction promoter and founder of one of the two cryogenic facilities operated today. Followers of that "movement" hope that their deep-frozen bodies will be resuscitated when modern medicine advances enough to treat whatever fatal diseases killed them. The process of cryopreservation ideally begins minutes after death. Large stainless steel vacuum-isolated containers store the body in minus 320 degrees Fahrenheit liquid nitrogen until the day of "re-animation". In the case of neurocryopreservation only the head is stored. The Alcor Society (Life Extension Foundation) was founded in 1972 in California. The only Cryonics movement is a nonprofit organization and performed in 1970 its first human cryopreservation service. In 1990 Alcor had grown to 300 members. In fear of earthquake risk Alcor moved its "patients" from California to Scottsdale, AZ. In the desert, "the perfect place for ice preservation," remaining units are rented to tenants. The resuscitation research is working on a prototype of an intermediate temperature storage device that tries to eliminate or reduce the risk of fracturing in cryopatients. Eighty-four patients and 872 living "cryonicists" already spent $150,000 to have their bodies scientifically preserved, and hope that someday they will be able to restore their bodies to full health. "They are convinced in a future simply worth coming back to," says Regina Pancake, Alcor's readiness coordinator. "Cryonicists always have hope, and are optimists who embrace the future instead of running from it", continues Regina. She is still impressed by Obama's inaugural speech, "about restoring science to its rightful place and wielding technology's wonders to raise healthcare qualities." Pancake is expecting that: "the world will be transformed by emerging technologies and thinks that people naturally become more optimistic that those technologies will work to eliminate diseases and will provide a way to slow aging and even death." The President's main inaugural sentence was e-mailed through the cryonics community in anticipation of finding new members (Magahern 2009). As of December 31, 2009, 1746 people have made cryonic arrangements. "Wealthy wired dudes" says Rudi Hoffmann, a fifty-three year-old cryonicist and certified financial planner. Rudi sells not wealthy but clever enough people life insurance policies to cover costs totaling $250,000 in addition to the transportation costs to Arizona and thus make the "dream" of cryopreservation possible (Theim 2010).

Up-to-date cryopreservation is very much requested. A nitrogen tank is booked for humans and beloved animals. But what happens when relatives and the cryogenic company are fighting over who gets the cash and head in probate. Mary Robbins, seventy-one, passed away on February 9, 2010. She left a $50,000 annuity to cover preservation costs for her head and brain. The daughter insists now that her mom changed her mind before she died. Now a court has to decide who will get the head (Turley 2010). Of course, journalists who are looking for sensational headlines will always raise the specter of charlatanism and thus the good science behind cryobiologists would be tainted in association. People forget that scientists generally pursue questions because they are curious. You have to keep in mind that: "It has become standard practice for scientists to sell their funding proposals through the technology that might arise out of their work," said Dr. Ken Muldrew from the University of Calgary, Alberta, Canada, in context with his textbook of the science of cryobiology. What happened to the people who were frozen in the late sixties and seventies? Unfortunately, the money required to keep them frozen did run out. Relatives stopped paying. No money no freezing. There was no need to address the ethical issue just because of a lack of proper provisions for the funds. Today cryonicist's have set up foundations to protect their interests. All that they ask of society is the freedom to do experiments on dead people who have given their consent to such experiments before dying (Muldrew 1999).

However current technology is far from being able to cryopreserve organs. The freezing process causes too much irreversible damage to ever be repaired. Cryobiology could be lifesaving, if not the freezing behavior of organ-cells, which build ice crystals, would injure the organ. Freezing draws water out of cells and ice crystal formation hinders the whole process. At present only cells and tissue can be reversible cryopreserved. It would be advantageous to freeze organs to have them available for organ transplantations. Unfortunately, the correct storage solution has not yet been discovered. Formation of ice crystals destroys the tissue. The only tissues able to survive cryo-freezing are embryos, egg-cells, adult stem cells, hematopoietic cells and umbilical cord cells (Sputtek 2007).

6.2 Qui s'excuse, s'accuse

Surplus created frozen embryos of in vitro fertilization clinics are stored in minus 82 Celsius nitrogen tanks, to await the cures of events. Couples debate with increasing frequency the fate of unwanted frozen embryos stored at fertility clinics in many countries. Courtney and Tim A. could not conceive their own child. They considered traditional adoption of a child but since they opposed stem cell research they ended up choosing embryo adoption. With the help of an adoption agency they received eleven frozen embryos.

Only three survived the thawing process. Courtney eventually gave birth to a boy on Sept. 15, 2003. Supporters of such a transfer argue that it enables couples to have children and prevents embryos from being destroyed for the sake of medical research. Biologists note that embryo transfer also causes

"Frozen" or adhesive stage

destruction because most of them will not come to term. Some ethicists consider that the phenomena of "embryo adoption" could create a market for embryos, as Jon Berkman wrote in a

2002 paper in the Scottish Journal of Theology. Courtney and her husband are very happy about the possibility they have had. "I think an embryo is a child, so it's hard for me to see someone donate it to embryonic stem cell research. My son and every baby is a miracle even though we saw this child in a petri dish," the couple told national TV. Her baby has two biological siblings from the couple who donated their extra embryos. "When the time comes her baby will get to know them," assures the mom. As of April 11, 2002, about 400,000 frozen embryos exist nationwide according to a survey by the Society of Assisted Reproduction Technology from Atlanta. Supporters and opponents of human embryonic stem cell research want to use them for their own goals, which is either for transfer or scientific work (Lieberman 2005).

"The use of embryonic stem cells in research is excused by purely scientific means. Following the road in this way can legalize everything," ponders Prof. Strauer, a famous German cardiologist (Strauer 2008). However, it is a very human trait to excuse oneself. In our days the individual and the masses are using a scientific excuse to pursue their research. Who wonders, since even Eve used the excuse and blamed the snake, after Adam accused Eve. The characteristics attributed to Eve happened long before the French countryman identified an excuse so very well with the idiom: "Who excuses himself, accuses himself" ("Qui s'excuse, s'accuse").

Prof. Robert Spaemann is a German philosopher, whose focus is on bioethics, ecology and human rights. He states in an article in the Frankfurter Allgemeine Zeitung from 1999 with the title:" The Teaching of a Good Cause", the following: "In 1952 the German Federal High Court condemned two physicians for abetting murder in the year 1941. At that time, the Nazi regime arranged massive euthanasia on mentally ill patients. The physicians supported the execution by registering patients into transfer lists, which meant that the patients were to be euthanized. The physicians cooperating in that selection process argued that with their participation they could spare more disabled patients. They arranged to save twenty five percent of disability patients' lives through that deal. Nevertheless, the German Federal High Court considered the physicians guilty in retrospect. The Christian moral teaching defines the dignity of a human person and does not consider its "selection". The Nazi ideology disregarded human dignity for certain whole groups of people. The Third Reich was a community of unethical values and laws. The Aryan race was considered superior. For tangible reasons the physicians complied with the immoral laws. The Court argued that the rule of choosing the lesser evil could not be applied to human beings. The behavior of the physicians was severe, because it was accompanied by killing persons whose humanity was denied. At that time, only a few physicians did not take part in the killing of mentally retarded people and rather preferred to lose their position and license rather than to participate in most indirect ways in this mass murder. Today it seems that Judeo-Christian values are being replaced by tolerance and relativism. Spaemann declared: "Nobody has the responsibility to intervene in something he cannot prevent taking place. Particularly not if he would have to do something unlawful to proceed. An action done in good faith depends not only on what I am doing. It also depends on the circumstances, the side effects, the alternatives and last but not least from the intention and motive of the acting person. There are certain things impossible to use not even for the most special purpose, because they are not for disposal" (Spaemann 1999).

Science is a product of human reason. In spite of this, we have to integrate scientific knowledge in the normative principles of the moral order. Throughout history ethics remains a key component in our understanding of how we should live in relationship to the material world. Even "Humanae Vitae" ("Of Human Life", an encyclical letter written by Pope Paul VI, 1969) points out the right intention, to use the sources of nature according to the principles of reason, and to transform them for the wellbeing of humans. "We still have to consider that environmental stewardship is rooted in the natural law," stated Pope Pius the XII in a speech addressed to Italian Catholic Union of Midwives on October 29, 1951 (Pius XII 1951). What are the consequences if humans intervene and disregard nature, for example, in order to determine their own fertility? Science is lacking comprehensive knowledge about pharmacokinetics and pharmacodynamics of hormones. They are aware of its benefits but know little about long term or adverse side effects. In spite of that it is immensely popular to implement hormones in order to attain a certain goal.

6.3 Chemical "liberation" of women

Oral Contraceptive Pills, known as the Birth Control Pill, are currently being used by over 10 million woman in the US. The anterior pituitary gland regulates several physiological processes, including stress, growth, and reproduction. The pituitary gland itself, controlled by the hypothalamus and by a negative feedback from the target organs. It also produces other hormones including Follicle-Stimulating Hormone (FSH) and Luteinizing hormone (LH). Each month, both hormones serve to stimulate the ovary to produce an egg, and to ovulate. The ovary produces the two central female hormones, estradiol, a type of estrogen, and progesterone, a type of progestin. Oral contraceptive pills are a combination of synthetic estrogen and progestin. They work by "telling" the pituitary gland to produce less FSH and LH resulting in the suppression, but not elimination of ovulation.

The 1920's were a time when hormones were an unexplored field in medicine. Ludwig Haberlandt is the pioneer who invented hormonal contraception in a female animal. Born in Graz, Austria he demonstrated temporary hormonal contraception by transplanting ovaries from a pregnant rabbit into non-pregnant animals (1922). In 1923 he began to highlight the importance of his research, with the result that his idea was seen as contradictory to the moral, ethic, religious and political agendas of the time in Europe. After he committed suicide, on July 22, 1932, the research ceased (Haberlandt 2009). The problem was that most hormones had not yet been chemically identified. The only choice was to extract them from animal organs, tainted with all the impurities and risks.

Those who know something about Russell E. Marker recall him as the "Father of the Pill". Marker, a chemist, never earned a Ph.D. His mentor claimed all of Marker's work to himself. The first African American scientist repeatedly tried to make clear that prevention of pregnancy, as a goal, never crossed his mind. His quest was to find a low-cost elixir in a new family of medicines called steroids. He never thought of creating anything like a social revolution. He just discovered something on a path to something else, far less ambitious than the final breakthrough, as he puts it (Asbell 1995). In the beginning of his professional career Marker worked at Ethyl Cooperation where he invented the octane rating system. In 1939 Russell E. Marker developed a method of synthesizing progesterone from a plant steroid, which unfortunately proved to be too expensive. In 1942, after three years of extensive research, Russell Marker, at that time a scientist at Penn State University, set out on a quest to manufacture exact replicas of human steroid hormones by a process of degeneration of similar molecules in plants. He found such a molecule in the Mexican wild yam (Mexican wild yam, like many natural remedies, enjoys a long history of use as a treatment for indigestion, coughs, morning sickness, gallbladder pain, menstrual cramps, contraception, joint pain, and nerve pain. The exact mechanism is not known, even today. The earliest use reaches back six or seven thousand years to Africa and Asia). When Marker brought back his new substance to the U.S. he found no one cared about what this "crazy scientist" had done. He could not find a single pharmaceutical company interested in developing his discovery. Unable to attract a research sponsor he founded his own company in Mexico, (Syntex), and produced "bio-identical" synthetic progesterone on a large scale (Web References to Professor R. Marker, 2010). This eventually led to the development of the combined oral contraceptive pill and a cheap ample supply of cortisone. Without his discovery and his efforts the birth control pill might have not been created. Thus you might say that the birth control pill was born unplanned.

Carl Djerassi is best known for his contribution to the development of the birth control pill. Djerassi was born in Vienna in 1923. After the outbreak of World War II, he emigrated to the U.S. In 1948, Carl enrolled and earned his Ph.D. at the University of Wisconsin. Unable to find a position in academia he worked four years in industry. In spring, 1949, Carl received a call from chemist George Rosenkranz, who invited him to head a research team at the small Mexican company called Syntex. This was an appealing job since steroids were the subject of Djerassi's, doctoral dissertation. He was very interested in their complicated structures and their prominent role in regulating physiological processes such as reproduction, digestion, calcium metabolism. In 1951 Djerassi's team synthesized cortisone and later on "norethisterone" from the Mexican yam. But the excitement was centered on the discovery of cortisone by using a readily available raw material, which could alleviate arthritis symptoms. Much less attention was given to the active ingredient "norethisterone" that was to become the active ingredient in the birth control pill. Syntex broke the monopoly of European pharmaceutical companies on steroid hormones. The price of progesterone dropped almost 200 fold in the coming years (Garfield 1983).

George Rosenkanz and Carl Djerassi established a pure gestagen to use as potential chemosterilants. Gestagen hormones are commonly used as contraceptives in women. Gestagens are similar to the hormone progesterone, which is abundantly present and the component hormone during pregnancy. Progesterone is secreted by the corpus luteum (develops from an ovarian follicle, following the release of the oocyte), adrenal cortex (situated along the perimeter of the adrenal gland), and placenta. Its chief function is to prepare the uterus for the reception and development of the fertilized ovum. It acts also as an anti-ovulatory agent. Estrogen belongs as well to the group of steroid compounds. It is produced in the developing follicles in the ovaries, the corpus luteum, and the placenta. Follicle-stimulating hormone (FSH) and luteinizing hormone (LH) stimulate the production of estrogen in the ovaries. Oral contraceptives contain a synthetic estrogen. Circulating in the blood it causes negative feedback and reduces the levels of FSH and LH.

The concept of the Pill is most often pinpointed to a New York dinner, arranged by Miss Margaret Sanger, a nurse and leading social activist in advocating birth control and woman's health, in 1951. Her guest was an ally in the birth control movement since the 1920's. Dr. Gregory Pincus, the foremost authority on the female component of fertility, had already accomplished the fertilization in a test tube of the eggs of a rabbit, participated with the illustrious circle. Unfortunately no one so far could reproduce the experiment independently. The proof of Pincus' claim did not occur and Dr. Gregory acquired fame only in being called "America's Count Frankenstein." Pincus struggled to found his private research laboratory, titled the Worcester Foundation for Experimental Biology. Margaret Sanger was able to win over Katharine McCormick, an extremely rich person that time. Katharine built dormitories, churches and hospitals and wanted to spend her money to support the National Birth Control Movement. Sanger's dream was as she wrote: "I consider that the world and almost our civilization for the next twenty-five years, is going to depend upon a simple, cheap, safe contraceptive to be used in poverty stricken slums, jungles and among the most ignorant people. I believe that now, immediately, there should be national sterilization for certain dysgenic types of our population who are being encouraged to breed and would die out were the government not feeding them." Sanger's friend McCormik persuaded Pincus with forty thousand dollars to produce a physiologically perfect oral contraceptive. Mrs. McCormick's investment soon expanded to two million dollars (Asbell 1995). The nurse Margaret Sanger and her friend Katharine McCormick, the second woman to earn a degree from Massachusetts Institute of Technology provided the drive for a mass development of the pill.

In 1956 the research team of Gregory Goodwin Pincus, John Rock and Celso-Ramon Gracia developed an estrogen-gestagen combination pill, which was tested for three years in the slums of San Juan in Puerto Rico, as such a test was illegal in Dr. Pincus' State of Massachusetts (Haumer 2008). In May 1960, Enovid was introduced to the US marketed by Searle Company. In 1961, Bayer Schering Pharma (then Schering) Berlin AG brought Anovlar to the European market. The hormone compound was fifty mcg of "Ethinyl Estradiol", and 4 mg of "Norethisterone" acetate per pill. Because of the high estrogen dose, the new medication could only be used for less than two years and under strict medical observations (Eheman a, 2000). The hormone dose has been gradually reduced since. The FDA approved Enovid 5 mg (5 mg norethynodrel and 75 µg mestranol) in July 1961. Nevertheless it was not available to married women in all States until 1965, and to unmarried women until 1972. It was argued that unmarried women did not "need" contraception. The new hormone preparation was only prescribed for married women if their husbands consented.

Estrogen and progesterone are responsible for the reproductive cycle regulation in all females in nature. A complex interplay of hormonal levels triggers the anterior pituitary gland. Under the influence of the hypothalamus, the anterior pituitary gland produces and secretes several peptide hormones that regulate many physiological processes including stress, growth and reproduction. Contraceptives finally demolish the precise interplay of hormones and physiological processes involving the hypothalamus, pituitary gland and ovaries. The strong intervention in the hormonal balance comes with significant side effects, which are thrombosis, pulmonary embolism, apoplexies, cardiac infarcts, cervical cancer, sterility, depression and a wide range of sexually transmitted diseases. Long-term consumption of contraceptives may significantly lessen the chances of conceiving later on in life. This is a very disillusioning state for females. In addition, a conception is not prevented in every case. It was estimated that contraceptives of the first generation failed in seven percent. Today we have the very probability that despite the pill the fertilization rate is 30 to 50 percent (Wloka 2000). The fallopian tubes are complex organs. The microenvironments in the different tubal segments favor the fertilization process and support early embryonic development. The tubal physiology permits the tubal transport of the embryo, which takes approximately five days. The inner cell lining of the uterus, the endometrium, slowly gets built up before ovulation and reaches its peak shortly after it. The blood flow increases to deliver oxygen and nutrients for the glandular cells. They enlarge now and contain important building blocks. The changes that take place in the endometrium during the second half of the menstrual cycle are seen as preparing the uterine lining for the nourishment and reception of the embryo. Upon arrival of the blastocyst in the uterus, the endometrium represents an optimal environment for implantation. On day seven and one-half to nine after conception, the trophoblast invades the uterus-endometrium to start the decidualization (Breburda et al. 2004, 2007).

Today women's monophasic and multiphasic contraceptives contain far less estrogen and progestin. A lower dose causes ovulation. High-tech ultrasound reveals that the rate of breakthrough ovulation in the first year is eleven percent but dramatically increased after that to forty four percent (Croxatto et al. 1982). Contraceptives also cause a markedly unfavorable decrease in the thickness of a woman's endometrial lining of fifty seven percent. A number of different research papers have studied this issue. It has been widely described in the medical literature because it concerns in vitro fertilization and shows that a newly conceived embryo is much less likely to implant on a thinner uterine lining (Schwartz et al. 1997).

Contraceptives might not prevent fertilization but rather the endometrium of the uterus is not prepared for an implantation. Consequently the embryo is starving when "nutrition" cannot be provided. Synchronous with implantation the gestation from the embryoblast is progressing. Contraceptives thus prevent the implantation in the mother's womb at the time the embryo reaches the blastocyst stage.

The abortifacient (an agent that causes an early abortion, specifically the death of the zygote, embryo or fetus after conception has occurred) nature of the pill was openly admitted by the most ardent pro-abortion supporters before the Supreme Court in 1989. The case Webster versus Reproductive Health Services received worldwide publicity. Mr. Frank Susman, who argued for the pro-abortion side told Justice Anthony Scalia: "The most common forms of what we generally call contraception act as abortifacients. They are correctly labeled as both" (The New York Times, 1989).

So far only the hormonal characteristics of contraceptives are addressed. Reproductive specialists consider contraceptives as highly potent pharmaceuticals with many unknown side effects, administered to healthy individuals without any medical indication (Ehemann b, 2000). The facts concerning contraceptives are not viewed as severe enough to alert pharmaceutical companies to take ovulation inhibitors off the market. The controversy of marketing contraceptives is not primarily a scientific issue whereas, it is more considered as a lack of perception. The lure of sex without consequences diminishes the awareness of potential risks. By far, immediate benefits exceed the fear of long term effects. It will take time until we feel the pain of this shortsighted act. Besides our body, we injure the environment. Instead of assuming responsibilities we develop a high level tolerance to enjoy dubious pleasure. The lack of self-discipline is justified by the expectation we have in using the advancements of modern medicine. We are pursuing happiness and are confident in the advances of modern medicine, even if it might harm us. The mentality of pleasure is assisting the contraceptive behavior and finally leads to the declining of the birth rate. Ultimately, we are responsible for having built up a society where only desired or even designed children are welcomed. The planned child generation is totally opposite to the postwar generation, which was characterized by making sacrifices and dealing with hardships. It is questionable whether a young person of the desired child generation, who is accustomed to getting everything immediately, will ever be able to handle difficult spiritual or physical life situations.

6.4 Environmental rights?

Condoms are considered by many as absolutely innocuous in matters of health issues. It is particularly seen as safe protection from sexually transmitted diseases, like HIV. One hardly knows about a study accomplished in 2006 at the Chemical and Veterinary Medicine State Laboratory in Stuttgart. The study showed that condoms release toxic and carcinogenic substances when exposed to an artificial sweat gland. A concentration of 660 microgram nitrosamines/per kilogram of rubber was set free revealing a sixty six times higher concentration than the maximally certified permitted dose from a baby-soother (Haumer b, 2000). It is not required that condoms are listed as carcinogenic nitrosamines. They also don't have warning labels about its biohazard substances. Nitrosamines are extremely hazardous substances, which appear in acid environment. Nitrites and amines are responsible for its formation. For example they can develop when heating cheese with bacon, or with smoking tobacco. The sperm and vaginal fluid are acid and promote the release of carcinogenic substances in condoms.

Pathological abnormalities of the male sexual organs are increasing. Already in 1993 they were twice as often diagnosed than thirty to fifty years earlier. The incidence of testicle cancer quadrupled in Europe and the United States. In the last thirty years the male fertility rate has constantly decreased. Environmentalists attribute these modifications to hormone pollution in drinking water. Synthetic estrogen "Ethinyl Estradiol" cannot be filtered from water and sewage treatment plants. They are passing through and are harming the ecosystem in unforeseen ways (Hügel and Süszmuth 2000).

Feminizing chemicals such as endocrine disruptors are making their way into our drinking water. Fish caught in Pittsburg rivers contained substances that mimic the actions of estrogen. Within their bodies chemicals from their habitat are

The feminized frogs

concentrated. Significant public health implications may occur. The high estrogen concentration in fish consumed by humans can cause estrogen sensitive breast cancer cells to grow (University of Pittsburg 2007). Contraceptives contaminate all water living creatures, like frogs, snails, and freshwater plankton.

Across the country male fish start to produce immature eggs in their sex organs, thus clearly feminizing. Hermaphrodite fish with both male and female characteristics are increasingly observed. Male and female fish develop liver and kidney malformations and slow down the reproductive cycle (Kidd 1994). Other estrogen mimicking compounds, for example alkylphenol ethoxylate surfactants (compounds used in pesticides, detergents and cosmetics) and the plasticizer, Bisphenol A (used in lacquers for dental treatment and to coat food cans and other metal containers) do not raise the same concern (Metclafe 2001).

Researchers began to investigate how low dose hormones can trigger health-concern effects. Pharmaceuticals like antibiotics, estrogen (natural and synthetic), steroids, prescription drugs, blood pressure reducers, hormones, psychiatric drugs, lipid regulators and pain killers remain in treated water that leaves the sewage plants and make their way into drinking water. Antibiotics and hormones used on livestock may also contribute to the problem. It is impossible to determine how prevalent they are because additional pharmaceutical waste is constantly added. Effects due to long term low dose exposure are not known at all. Contaminants in soil and water may build up antibiotic resistances. This situation is considered an extreme paradox, because the protecting safeguards of bacteria and toxins suddenly harm us. The pharmacological activity in relation of its toxicity differs in mammalian and humans. Medications have a more diverse profile of effects in nature than they show in humans. For example, selective serotonin reuptake inhibitors are taken by humans to avoid depression. In shellfish and nearly all other aquatic animals, it affects the premature release of larvae.

Fish, Daphnia (water fleas) and algae are excellent indicators for the ecotoxicological and toxicological effects of groundwater contaminants on aqueous ecosystems, groundwater-ecosystems and groundwater-dependent terrestrial ecosystems. Ethinyl Estradiol should prevent pregnancy or is prescribed for menopausal symptoms, whereas a nano-gram concentration of it in one liter of water is already affecting fish. The varying components of the medical cabinet will even intensify the maximum toxic effects when mingled together (Kaspar 2005). Studies will determine whether a chemical communication exists between the endocrine modulators of the human body and the epidemic of chemicals in drinking water. Beside the birth control pill, scientists detected in our water resources food additives, some skin cream, pharmaceuticals and personal care products. Many of these chemicals have proven to mimic or inhibit the hormonal balance in the human body and could be a threat for female reproduction. Infertility is also very damaging to the livestock animal breeding. Neoplasia in hormone producing organs is increasingly recorded. In addition, behavioral disorders were observed in newborns (Koroljow 2007). Whitehead eagles have habitat requirements that must be met in order to successfully reproduce and survive during the winter or non-nesting season. In 1952, Florida reported fertility problems observed in bald eagles. It was supposed that environmental contaminants influenced the hormone system (Broley 1958). Silver sea gulls showed abnormalities during embryonic development, which resembled hormone imbalance (Gilbertson et al. 1995). Reproductive disorders in humans and wild animals caused by low-level estrogen substances are mentioned frequently (Koroljow 2007). Estrogen of the anti-baby pills provide an example of a dramatic shift in ecosystem conditions, with unforeseen effects in animals and the potential to impact human health.

7 Therapeutic imperative- conservation of all creatures

The Cosmos Journal first coined the term Bioethic in an article in 1927. It refers to all areas of gene-ethics, animal-ethics, environmental-ethics and neuro-ethics. Bioethics is embedded in a system of moral principles, standards and codes of conduct governing what society considers worthy to protect (Reiter 2002). The ethical challenge is to balance different risks, benefits and duties. Concepts of bioethics are present in literature, art, music, culture, philosophy and religion. It is predictive probability that somebody who loves animals becomes an animal-rights supporter and activist. He will do everything he can to protect animals from any cruelty.

However, in some countries greed for profit is more important than fundamental animal rights. Therefore, it seems common to neglect commercial standard animal husbandry, in favor of economic competition. Barbaric grievances are tolerated with the apology that it would serve the growing affluence of the country. The western consumer's attitude and desire toward cheap, but high-ranking luxury products cause cruelty to animals. Human interest should not be the "excuse" for animals to suffer! In 1999, Switzerland's upper house of parliament legally represented the opinion that animals cannot be longer defined as an item. Consequently, in 2003, a new law established that animals have to be respected as living beings, created alongside with humans.

Paragraph one, of the "German Animal Protection Law" views animals as co-creation. Humans are responsible for the well-being of animals.

By law, no one is allowed to harm an animal without a good reason. It is imperative to show mercy towards all living beings. This prospect includes nature in its entirety.

Species-appropriate husbandry?

Furthermore, the rational view matches the biblical vision of stewardship (Reiter 2003). Evidently countries that do not respect animal rights also do not protect human embryos. Besides, what moral status does a human embryo have? For many, an embryo is not a human, not an animal, but also not an item. To kill a baby after birth is considered more severe than to take human life from an embryo or fetus. On the contrary, pregnant women who are smoking are viewed irresponsible, because they risk injuring the unborn child, whereas people are not scandalized as much by passive smoking (Warkus 2005).

7.1 Growing demand for medical technology in aging societies

What are the criteria for speaking about life? The cell is the smallest structural and functional unit of all known living organisms. The word cell comes from Latin, and means small room. Bacteria are unicellular and thus consist of a single cell. All other organisms are multi-cellular. Each cell contains its hereditary information, which qualifies it to regulate cell function in order to establish a well-organized vital system. The genetic code is inherited from the preexisting cells and is transmitted to the next generation. The general principle of a cell is self-maintenance. It takes up nutrients and converts them into energy in its own metabolism. The chemical energy is used in a unique metabolic pathway to carry out specialized functions and cell division if necessary. A human embryo has its origin also in a single cell called zygote. It contains a human DNA, responsible for the proper development. Prokaryotic cells are single cells like bacteria. They lack a cell nucleus and have the cell genome DNA for their long-term information storage located in the cytoplasmic region. The biological information gives the bacteria the title. The mycobacterium tuberculosis is a gram-positive pathogenic bacteria which causes tuberculosis in humans. Mycobacterium bovis is usually pathogenic for cattle. It caused tuberculosis in animals long before invading humans. The DNA of the bacteria specifies the disease. No one would change the nomenclature. In contrast, nobody would treat a tuberculosis bacterium without the necessary precautions to prevent transmissions. Elderly people might be more familiar with another, more delightful gram-positive bacteria, namely clostridium botulinum. Botulinum means sausage, but the pleasing affiliation is not originated by meat-products. Botulinus is well known to aging people who wish to regain youthful attractiveness. For sure, they know botox, the toxin of botulinum clostridium bacterium. It is an obligatory anaerobe organism able to form endospores that allow them to survive very long. They were isolated from home cured ham and sausage links or canned food such as green beans. Botulin toxin produced by clostridium botulinum is the most dangerous neurotoxin. It takes about 75 Nano grams to kill a person. 500 grams would kill the entire human population. It works by blocking acetylcholine receptors on the muscle side. Facial wrinkles are caused by muscle contraction. Botox is used to selectively paralyze muscles with the effect of a temporary relief of wrinkles. It has other medical purposes as well such as treating severe facial pain. Beauty surgeons convert the high potential toxin into cash. Botox is recognized as a fountain of youth, or an alternative way against the annoyance of aging, despite the fact that the age will be the same with or without botox. A very expensive treatment will make it at least impossible for the patients to laugh or cry for the next three months. What are ordinary people doing to obtain the elusive goal of eternal youth?

Uncontrollable risks and uninvestigated undesirable side effects do not prevent people from paying huge amounts of money to perceive an improvement in their physical appearance. In this view, the costly treatment more available to the wealthy creates social classes. The ones to whom cosmetic care is affordable, in contrast to the ones who cannot even afford basic health care. But perhaps there is no reason to be envious of the rich, as they accept unknown risks for their expensive cosmetic treatments.

In 2008 Dr. Caleo from Pisa showed in a clinical trial that botox was found far away from injection sites. Botox potentially migrates inversely to the direction of nerve pathways. Three days after its application botox was found in the brain stem, where it blocked the contra-lateral non-treated body side. It was also found to migrate to the retina synapses when injected into peri-orbital wrinkles. Currently only data from animal trials, and no human data, are available in regard to the remote effects of botox. Therefore the FDA issued special warnings and recommends that the treatment with botox be done only by experienced hands. The therapist can lose control when the toxin is migrating. For example treating sweat glands in order to reduce secretion can be a health hazard. However, for children with muscle disorders, botox shows to be a beneficial medication. Nevertheless, its use decreases because of the discovered side effects, particularly the risk of sudden respiratory arrest. Thus, the result of Pisa is changing pediatric treatment regimens (Antonucci 2008).

Pisa is well known for its famous leaning tower. One of the most popular biblical narrations in the Old Testament also refers to a tower, the tower of Babel. It is featured in chapter 10 and 11 of the book of Genesis. The construction of the tower of Babel was undertaken as an attempt to approach JAHWE. The pride of humans and their ambition was aimed at the purpose of reaching heaven. God stopped this endeavor. Genesis 11,5-8 states: "And the Lord came down to see the town and the tower which the children of Adam were building. The LORD said, "Behold, they are one people, and they all have the same language. And this is what they began to do, and now nothing, which they intend to do will be impossible for them. Come, let us go down and take away the meaning of their language, so that they will not be able to make themselves clear to one another. So the LORD scattered them abroad from there over the face of the whole earth; and they stopped building the city." Babel is a symbol for the evolution of the various human tongues. Indeed, it is quite amazing to this day, that linguists are uncertain about the development of human language. The human language seems to be a unique phenomenon without significant analogues in the animal world. Noam Chomsky, a prominent linguist stated that: "Animal signals are fixed, limited, and instinctively used. On the contrary human communication is very variable, unlimited and freely used" (Chomsky 1972). Linguists are not able to state expressly where humans got the skills to know how to speak or whence its genesis comes. Children who grow up without verbal communication cannot speak (Crystal 2002). Religious people are convinced that only God can create matter, energy and life from nothing, consequently only God can create language out of nothing. Scientists cannot explain why man has only the instinct to learn language, but is not able to create it. A child must first be exposed to the sounds of words to use language. Then it can instinctively abstract grammatical rules of a language and apply them creatively. To think that "man, just advancing out of bestiality, could create language is absurd." even for theologians like Warkulwiz. Indo-Germanic, the language spoken before 6000 years was extraordinarily rich in expressions and had a phenomenal grammatical wealth. In the course of thousands of years, the archaic language lost its comprehensive terms. Our so-called high-level languages cannot be compared to the protolanguage. "As life comes only from life, so language comes only from language. God specially created the original languages and infused them into the minds of the first users" (Warkulwiz 2007). "In principio erat Verbum, et Verbum erat apud Deum et Deus erat Verbum" (John 1:1). The first verse in the Gospel of John reads, "In the beginning was the Word, and the Word was with God, and the Word was God."

7.2 Eternal youth

People want to fight health challenges that accompany aging. The years pass but they still want to look young, healthy and vital. They are sick of all the plagues that make their lives a misery. Desperately they search for a panacea to get the youthful glow back. Some decided to find a solution in natural remedies. Stay forever young is the message of a full range of products that promise to improve the general well-being. Spot ads, "support a healthy vision with products unavailable anywhere else and you will gain eternal youth," are constantly hammering our minds. The ancient Aztecs or Mayan already knew about the treasury of herbs. Especially about the advantages of Phytoestrogen's, which have the properties of estrogenic and rejuvenating effects, mostly to the female body. What you eat can have a direct impact on your brain chemistry, can affect your hormones, energy and stress level. Even so the extent of psychological versus biological influence is debatable and the lines blurred. Most of the remarkable health benefits in fruits and vegetables are still unidentified. But there are companies out there who are using scientific tests that accurately measure your body's biochemistry. Then the researchers recommend what to eat to lead you back to health. The cure, connected to the whole foods store advises you to keep, for example, copper out of your diet, because you don't need to experience: "frontal headaches, depression, fatigue, emotional volatility, weight gain and food craving, constipation." Besides that, copper also helps estrogen levels to remain unusually high. Because copper is a primary component of the pill, it helps to control the ovulation. "It is more and more observed that our generation gets so toxic with copper," states Beth Ellen Rosenthal in her article for the Natural Health and Longevity Resource Center. Beth Ellen knows a lot about estrogen. She was under its influence from the moment her mom conceived her. Her tale is that her mom desperately tried to have a baby after two miscarriages. Her mom finally participated in a special study at the University of Chicago to test a promising new drug DES. Fifty-two injections of that artificial form of estrogen during pregnancy would allow her to carry a baby to term. Later researchers discovered that babies with massive exposure to DES, (diethylstilbestrol) a synthetic estrogen, during embryonic development had severe deformities in the sexual organs of both males and females. Beth does not even want to know what researchers will find in twenty years from now about estrogen replacement therapy. She herself spent her lifetime coping with the "wonder" drug and certainly did not want to be exposed more than necessary to external sources of estrogen, hidden in dairy products and beef that were commercially raised. Besides, external sources of estrogen led people to purchase beef that is naturally grass-fed with no hormones (Rosenthal).

In September 2009, a study in rats was published in the Journal Toxicological Sciences that revealed that male offspring rats established mal-development of their sex organs and showed a loss of fertility when the rat mother was under the influence of estrogen during pregnancy. The estrogen exposure rat experiment (low amount of estrogen, as a woman may use daily in contraceptives) is relevant to humans because more than 50 million woman worldwide take contraceptive pills. In one part of the rat study the main estrogen in birth control pills was compared with the synthetic estrogen DES. In earlier times it was common to administer DES to pregnant women to help control morning sickness. Today it is known that DES is a teratogen. The results showed that the kind of estrogen found in oral contraceptives is as toxic to male rat reproductive organs as is DES. However, estrogenic teratogen diethylstilbestrol (DES) was banned for use as a pharmaceutical in the early 1970's. It caused rare cancer and reproductive deformities in the children of the mothers who took the drug (Mathews et al. 2009). The recent animal research suggests that women who continue to take birth control pills well into pregnancy may skew their sons' reproductive development and may diminish their boys' fertility. "At least three to four percent may take the pill into the second trimester," underlines the expert Michele a La Merrill to Environmental Health News. It is further feared that certain pesticides and industrial chemicals that act like estrogen may have the same effects to the reproductive organs and fertility of the offspring (La Merrill 2010).

Is the anti-aging-idea all about business, able to offend a "tree-hugger"? Potentially it is beyond the capabilities of science to free humans from physical aging. Might it only remain a popular theme in trivial fictional novels and advertisements? Several fountains of youth have been recently pursued. One still remembers the "Living Cell Therapy" praised as optimal therapy and the uttermost opportunity to regenerate in a completely natural way. Scientific studies showed that biological aging and degeneration begins at thirty. It has been noted that in our hectic times high blood pressure and stress consume our last energy reserves. The living cell treatment promised to regain the life standard and health of younger years. It was also considered to be prophylactic for a variety of health problems. People who wished to strengthen the body's defense mechanism, like in case of allergies, also used living cells. Human beings desire to look younger and warnings of dubious treatments are disregarded. It is an age-old dream of mortal persons to avoid the detrimental effects of aging. It seems to be one of the greatest requests of living souls, handed on from generation to generation. The public tends to trust blindly the promises of anti-aging. "Living Cell Therapy" was one of the most popular contemporary therapies in the eighties and nineties. It consisted of embryonic cell suspensions from unborn or juvenile animals like calves, sheep-lambs, or goats, injected subcutaneously to patients. However, living cell therapies quickly lost its market attractiveness, due to the possible danger of transmitting diseases from animals to humans. Adults were also put at risk to suffer an allergic shock. In 1997, "Living Cell Therapies" was abolished in Germany after several reported deaths. Three years later the prohibition was again lifted and justified by some beneficial evidence.

7.3 Regenerative strategy

Today stem cell research seems to replace the promises of the "Living Cell Therapy". Nowadays efforts are made with living cells from bone marrow. Adult stem cells were able to regenerate or develop new nerve cells. Experiments on mice showed that intravenously applied bone marrow cells developed to brain cells. Thus the donor-derived cells are nothing more than progenitors for brain cells. Scientists were utterly astonished that the marrow cells found exactly the injured cell type out of approximately 10,000 different nerve cells they were supposed to repair (Brazelton et al. 2000). However, scientists admit that they still have a very incomplete understanding of the functions of brain cells. Virtually absent is the substantial knowledge of how undifferentiated embryonic stem cells will react when administered to the body. "The secret of the pathway of differentiation might perhaps be solved within the next two decades" (Bogdahn 2000).

To achieve an increasing recruitment of progenitor cells from bone marrow after myocardial infarct is the research-topic of the Cardiovascular Division and Department of Medicine at Brigham and Woman's Hospital and Harvard Medical School in Boston, MA. Dr. Steinhauser and his team describe in the issue of Cell Stem Cell from April, 2009, a pharmaco-regenerative strategy that improved heart function after failure. Adult cardiac precursor populations are present in bone marrow and there are opportunities to harness the endogenous regenerative potential without exogenous delivery of stem cells. The granulocyte colony stimulating factor (G-CSF) has stem cell mobilizing properties which are activated after myocardial infarct. The author of the paper stimulated the progenitor recruitment for a longer time than the body usually does. In addition SDF-1 (chemokine, stromal cell-derived factor, type I), a homing signal, directed the cells to the site of injury. The injury is ischemic and the production of SDF-1 is stimulated by hypoxia. It functions as a chemo attractant for any precursor cells. The combination of G-SCF and SDF-1 ultimately led to the generation of new blood vessels, improved cardiac function and increased survival. G-CSF is widely used in transplantation protocols to mobilize hematopoietic stem cells from the bone marrow into circulation prior to peripheral harvesting. Endogenous adult stem cell progenitors were recruited to injured tissue without employing an exogenously prepared cell preparation and without their invasive delivery (Steinhauser and Lee 2009).

"Chemical as well as mechanical stimulation likely guides the differentiation of adult stem cells into many cells in all the correct places," explained Dr. Doris Taylor, stem cell researcher from the University of Minnesota recently in an article of "New Scientist". The expert is referring to the repopulation of a decellularized (stripped-down) rat heart scaffold. Tiny thumping hearts sit in her lab, hooked up to an artificial blood supply. The rat hearts beat just as if they were inside a live animal. Human organs for transplants are scarce. Traditional xenotransplants from animals provoke a catastrophic immune response. In the United States about

3,000 individuals await a donor heart. Twenty-two million are living with heart failure. A bio-artificial heart would be a theoretical alternative. But the generation of such an organ requires engineering of cardiac architecture, developing the appropriate cellular constituents and pump function (Ott et al. 2008). Taylor sees in her method a limitless supply for organ transplants. All she needs to do after preserving an acellular heart frame is to let the added cells grow naturally. "It is letting nature do most of the work", Dr. Taylor says. Adult stem cells that "clothe" the naked scaffold are taken from another rat's heart. Thus the organ stands a higher chance of being accepted by the immune system. Growth factors and peptides signal to stem cells how many should migrate and to which areas. Dr. Taylor now wants to repeat the achievement on a much larger scale. She will "decellularize" hearts, livers and other organs taken either from human cadavers or from larger animals such as pigs, and coat them in stem cells harvested from people. The idea is to take an organ from a human donor or animal and use a mild detergent to strip away flesh cells and DNA. Left over is the collagen scaffold an "immunologically inert" protein. Afterwards stem cells from the relevant patient are added. They will differentiate into all the cells the organ needs for the most adequate function. The concept worked already to "replace" simple organs like a trachea (windpipe) donated from a dead donor. "All the cells need is oxygen. It was a big challenge with complex organs to ensure that all their cells are infused with blood. Pumping blood and nutrients through the organ produces pressure and is helping to determine how cells differentiate there," suggested Dr. Taylor. "The next step is to see if the transplants are replacing existing heart and keeping the animal alive and healthy," pointed out the adult stem cell expert. She casually mentioned the difficulties. Some find the idea of animal parts unacceptable, off-putting and dangerous. These folks forget that millions of patients have already decellularized pig parts without being infected by stowaway pig viruses or do not suffer from disastrous immunological reactions. For example new pig heart valves, or pig gut to produce patches that help the healing of diabetic ulcers, hernias and strained ligaments. The "spare parts" have been thoroughly decellularized and it is not any more necessary for the patients to take immunosuppressive drugs. Chris Mason, professor in regenerative medicine at the University College of London is convinced that one day we will implant complicated pig organs into people (Coghlan 2009).

7.4 Fully informed consent

The research community is split, even so all scientists try to achieve the same goals by desperately searching for the possibility of having enough organs for transplants. The broad types of stem cells that exist are: adult stem cells, found in adult organisms and act as progenitor cells to repair systems for the body; and embryonic stem cells, isolated by destroying the inner cell mass, also called embryoblast, of the blastocyst staged developing embryo. Human embryonic stem cells are considered as raw material for biomedicine. Human embryonic stem cells are not legally protected therefore scientists claim unlimited access. Scientists have no definite knowledge about their potential or long-range therapeutic benefits. It is very much hoped that with embryonic stem cells a new therapeutic area and limitless cures are possible. However many skeptical voices warn for example about carcinogenic effects of embryonic stem cells. A substantial deficit of knowledge and moral concerns regarding embryonic stem cells overshadows the euphoria.

Human rights have an established place in biomedical research. The European convention on human rights and biomedicine stated fundamental requirements for fully informed consent about the experimental nature of the study. Patients have the power to exercise their freedom of choice to participate in research. Research involving human subjects can only be carried out after meeting the following conditions:

Considered unable to give informed consent!

There are no alternatives of comparable effectiveness to research on humans. The risks, which may be incurred by that person, are not disproportionate to the potential benefits of the research. An institutional ethical review board proofed the study. The subjects, undergoing research have been fully informed about their rights. The safeguard prescribed the law for their protection. The patient giving consent may freely withdraw at any time. Biomedical research is built on principles protecting human rights and dignity of the participating persons. Informed consent is fundamental, but does not give the investigators the right to act without restriction. The neglect to protect a vulnerable group of humans and those who would not have been able or were never asked to consent to a research trial is still burdening the memory of some physicians especially in Germany during the Nazi area. Results from so-called scientific trails in concentration camps are thus ethically unacceptable and are considered criminal.

Ethical principles are established in medical research and assist in dilemmas between personal human rights and scientific development. This raises the question to whom is humankind morally responsible? Is my neighbor or my ancestor responsible for setting up moral rules? At least all humans are considered equal before the law. Principles should be applied universally. Therefore, nobody can use another person as a means to his own end. In addition, everybody is morally responsible for his own actions. Moral principles cannot be defined by individuals or governing bodies, as they would become subjective and manipulative. Immanuel Kant, a German philosopher, was one of the most influential thinkers of modern Europe. He stated that moral principles must come from God. That is manifested to us through the law of human nature, also called the natural law, which is the proximate source of all obligations (Fagothey c, 1958). Thus the moral law finally expresses itself in the natural law, best defined in the Decalogue, the ten commandments or Mosaic Law, which finest describes the inner voice of human conscience naturally found in every person. In 1944, Pope Pius XII stated that each human life gets its value and dignity from God and not from any human authority. Embryonic life has the same value as human life in its advanced stages. Thus destroying human life at any stage is to be considered an intrinsic evil. The act is not less corrupt when destroying an embryo rather than an older person. In every case innocent life is taken (Reiter 2002). The Pope stated that during the Nazi dictatorship, a time when it was common to use justifications like unprofitable eater (unnützer Esser), unworthy life (unwertes Leben), inferior race (Untermensch). Also at that time, a new law was announced to prevent inherited diseases in newborn children (Gesetz zur Verhütung erbkranken Nachwuchses). The consequence of the state-ordered discrimination of congenital disorders or handicapped humans resulted in mass murdering. During that time, German genetic scientists were considerably involved in these so-called euthanasia programs. By current genetic standards the demanded eugenic cleansing is unreasonable, even by its own immoral logic, as carriers of autosomal recessive hereditary disorders were not identifiable years ago. At the international meeting of genetics in 2008 in Berlin, German leading researchers still admit to a historical heavy guilt incurred during the Nazi time.

Genetic determinism is generally accepted in many scientific circles. Many factors contribute to the phenotype of a certain disease. A genetic predisposition may contribute to the likelihood of developing a disorder. Environmental triggers, lifestyle and epigenes are what actually make the individual sick. As Mr. Femi Soremekun, Managing Director of a Pharmaceutical company explained: "Genes and culture load the gun, but the economic environment pulls the trigger." A recent article in Time (18 Jan, 2010) stated that: "The new field of epigenetics is showing how your environment and your choices can influence your genetic code and that of your kids" (p. 49). That means, powerful environmental conditions leave an imprint on the genetic material that can affect one's offspring. Genes do not change, but epigenetic marks on top of the gene "tell" the genes to switch off or on. Scientists explain, "If the gene is the hardware, then the epigene is the software."

That is, "you're going to have the same chip in there, the same genome, but different software. And the outcome is a different cell type" (Geisler 2010). Disregarding these facts will overestimate genetic diagnostic testing performed on unborn or embryos before implantation. Every fifth or sixth pregnant woman is taking advantage of a genetic screening. It is feared that pre-born humans with genetic abnormalities will have to bear the consequences. Nevertheless the prenatal diagnostic is already called "grass roots eugenic", because ninety percent of pregnancies are terminated in case of a pathological chromosomal aberration like for example Down syndrome or Trisomy 21 (Bonfranchi 1996). In 2008 Susannah Baruch from John Hopkins University stated that genetic therapy would be an alternative to abortion. Since 1990, science tries to help patients with serious genetically caused diseases to get healthy children. For that purpose harmless viruses loaded with the genes that are able to replace the defective genes were transferred into sperm and egg cells. However the interaction of the genes themselves is very complex and it might lead to some surprises. Above that, it is very difficult to integrate the foreign gene at the correct place. Reproductive scientists hope that in the next 10 years the technology will be ready for such procedures. The benefits for the therapy remain questionable (Schulte von Drach 2008).

Human life begins with a zygote. In this regard, all forms of life arise equally. Do we admit that human life has a certain dignity and value from its beginning? Kant argued that the dignity of man is based on common sense. Every human being belongs to the human family and thus has the right to life, freedom and personal dignity.

In 1996, the Convention for the Protection of Human Rights and Fundamental Freedoms (also called the "European Convention on Human Rights") stated in article 18 that the creation of human embryos for research purposes is not permissible. Since 1991, the embryo protection law forbids genetically encroachments (Reiter 2002). The respect for life should characterize every civilized nation. Crimes against humanity were perpetrated in Nazi Germany. Some proponents of embryonic stem cell research consider research on surplus embryos permissible, as they otherwise will be eliminated anyway. They declare embryos as unwanted and thus suggest that the value of an embryo is determined by someone else whether it is wanted and if not it can be eliminated and consequently used by scientists for experimentation (Willke 2006). Nevertheless this constitutes the abuse of human embryos. Frozen human beings at the beginning of life are in a situation which essentially is not different from that of other living humans who are spending their last years in nursing homes. Because of senility, Alzheimer's disease or other debilitating conditions they might be also completely unaware of the world around them. They too are "only going to die anyway", but no one proposes to use them for medical research or to extract body parts in order to enhance the life or health of others (Flader 2002).

7.5 Alzheimer, a late onset from Down syndrome

Nearly twenty years ago Dr. Huntington Potter started to postulate that Alzheimer and Down syndrome were the same disease. In 2010 Dr. Potter provided scientific proof that he was right. A study from the University in Tampa Florida implicates that Down syndrome, Alzheimer, artery-clogging cardiovascular disease and even diabetes appear to share a common disease mechanism. Experts have definitely shown that Alzheimer is linked with Down's syndrome (Baier 2010).

Scientists do not know what causes Alzheimer brains to accumulate plaques (primary made of beta protein), but they developed some theories. In November, 2010, Lenzken and his team from the University of Milano, Italy speculated that mitochondrial damage and dysfunction play a role in a number of neurodegenerative disorders like Alzheimer and Parkinson, or Lateral Sclerosis (Lenzken et al. 2010). Besides a toxic exposure or infectious agent theory, the genetic theory seems to be quite important. The genetic theory relies on the discovery that there is a gene on chromosome 21, which produces the beta protein, found in Alzheimer Plaques. However, other researchers consider that the gene itself is not the main factor for Alzheimer, but might be a link in the chain to explain what causes the disorder. They argue that if a genetic factor causes the disease, genetically identical twins would both be affected by Alzheimer's. Many cases have been found where only one twin has the disease. Thus environmental factors seem to play the more prominent role (St George-Hyslop PH 2000). Nevertheless, a study in 2010 done in 2 Italian families described that a mutation, known as London mutation, of the protein gene on chromosome 21 is associated with Alzheimer's. Out of seven (2) siblings, three (2) carriers of the mutation were diagnosed with Alzheimer's. One sibling with the same mutation has only shown signs of executive dysfunction so far (Talarico et al. 2010). In all of these patients, the mutation leads to the abundant amount of the beta protein. Research tries to prevent or slow down Alzheimer's disease by providing a way to decrease beta proteins.

Many scientists believe that the majority of sporadic (non-inherited) cases of Alzheimer's disease is much more common. The German physician Alois Alzheimer, after whom the disorder is named, first described the disease 1907. Until the 1960's and 1970's it was believed that Alzheimer is a quite rare neurological disorder, whereas today it is more common to diagnose all elderly dementing people as "Alzheimer's". In the patient's brain several different changes occur. Experts observe "neurofibrillary tangles" and "senile plaques" under the microscope at autopsy. Neurofibrillary tangles are fibers, which probably disrupt the normal function of nerve cells. In a senile plaque parts of the nerve cells degenerate and are substituted by a substance called "amyloid". Amyloid is a protein that does not play a role in the human physiology. Nobody can explain for sure why it forms. The plaques also contain aluminum and silicon. Some researchers think this causes the disease. While scientists are still trying to determine what causes Alzheimer's. The severity of the disorder is closely related to the number of plaques and tangles, but also it depends on which areas are greatly affected. In the early stage of Alzheimer's areas associated with learning and memory are covered with the "foreign" objects. It explains the severe deterioration in memory and self-care skills that can cause distressing changes in personality. Family members will notice the difficulty of the loved one in remembering new things. Later, one suffers also from becoming suspicious, irritable, changeable in mood, stubborn, impatient or easily upset. In later stages all aspects of intelligence are affected and the patient may even may be unable to communicate in a meaningful way, explains Dr. A. Jorm from the National Health & Medical Research Council Social Psychiatry Research Unit Australian National University in his internet article published by Scribd: "Alzheimer's, what research has taught us about Alzheimer's and Dementia Disease."

In January, 2010, Dr. Potter and colleagues at the Florida Alzheimer's Disease Research Center, USF Health Byrd Alzheimer's Institute reported in Molecular Biology of the Cell and in PLoS One, that the beta amyloid (amyloid protein) associated with Alzheimer's damage the microtubule transport system responsible to move chromosomes inside the cells. The microtubules are segregating newly duplicated chromosomes as cells divide. When the network is disrupted new cells have the wrong number of chromosomes and an abnormal assortment of genes. Potter showed that Alzheimer's patients harbor some cells with three copies of chromosome 21, known as trisomy 21. People with Down syndrome share the same characteristics with three copies of the beta amyloid gene on chromosome 21. Scientists consider Alzheimer as a late onset from Down syndrome. The presence of an extra chromosome leads to an overexpression of the genes located in that chromosome. For that reason the brain pathology of a thirty to forty years-old carrier of Down syndrome is identical to Alzheimer's. Dr. Potter reported: "Beta amyloid basically creates potholes in the protein highways that move cargo including chromosomes, around inside cells. Alzheimer disease is probably caused in part from the continuous development of new trisomy 21 nerve cells, which amplify the disease process by producing extra beta amyloid." In addition many Alzheimer's disease patients develop vascular diseases and diabetes. Dr. Potter explained that several key proteins- including insulin receptors and receptors for the brain signaling molecules- are also likely locked inside cells. But with Alzheimer's the transport system is damaged by amyloid or other factors. For the expert, Alzheimer's, cardiovascular diseases and diabetes cannot be longer described as independent conditions, because they develop in the same patient (Granic et al. 2010 and Abisambra et al. 2010).

Are people with Down syndrome at risk to develop Alzheimer? Very early in life People with Down syndrome develop brain plaques and tangles as common in Alzheimer's disease. Scientist Elizabeth Head of the University of Kentucky's Sanders-Brown Center on Aging observed, that Down

Syndrome Patients repair and compensate the damage. "Their brains may be clearing the plaques." Dr. Head started to recruit Down syndrome patients for a study on biomarkers of Alzheimer's. Head reported that in aged Down syndrome patients the protective process slows down. However, not everybody succumbs and will automatically develop dementia.

Doctors don't know how an extra copy of chromosome 21 causes or prevent diseases. They assume that getting a larger dose of a gene affects the body's susceptibility to a disease, whereas the unique genetic profile also protects them from many common ailments. Down syndrome carriers might be overweight with high cholesterol but they never develop high blood pressure. They are having fewer tumors and rarely develop macular degeneration. Genes on Chromosome 21 inhibit the growth of blood vessels necessary for tumor growth and responsible for macular degeneration. Researchers are already trying to develop anti-cancer treatments based on genes found on chromosome 21, said Roger Reeves of John Hopkins School of Medicine. Researchers are grateful to the Down syndrome community for teaching scientists so much (Szabo 2010).

Might a child born today reasonably expect that a cure for Alzheimer's will be found, as they get older? In 2002, fifteen embryos were tested for inherited early-onset genes related to Alzheimer's disease. The selected mutation free embryo was transferred, yielding a clinical pregnancy and birth of a "healthy" child, which was free of the predisposing gene mutation as Yury Verlinsky reported in the Journal of the American Medical Association on February 27, 2002. As we know in 2010, Alzheimer is seldom inherited, besides a sole gene causes major changes but they work together with lifestyle factors, epigenes and are influenced by the environment. The human genome has just fewer than 20,000 genes. It is difficult to predict how many of these will influence the risk of developing Alzheimer.

David Shenk called one of his books <u>The Paradoxical Nature of Information Technology</u>. He has sympathy with the idea that we all want a world without Down's, Alzheimer's and Huntington's. "But when the vaccine against these disorders takes the forms of genetic knowledge and when that knowledge comes with a sneak preview of the full catalogue of weaknesses in each of us, solutions start to look like potential problems. With the early peek comes a transfer of control from natural law to human law."

In December, 1997, David Shenk refers to the "if - then conclusions", in his essay in Harper's Magazine: "Biocapitalism: What price the genetic revolution?" If the child is defective will it be kept? We are pursuing the human genome for the as we call it "good" reason -to abort the child with Down syndrome, or to select the healthy embryo after we got the genetic snapshot. "Should prospective parents who want a child be allowed to refuse a particular type of child?" asked Shenk. This includes a substantial number of false positives, which have no problems at all. Shenk himself was told his wife would deliver a baby with Down syndrome. He and his wife know that the amniocentesis procedure itself could lead to a spontaneous miscarriage. He did not want to take that chance only for the purpose to discover for sure his daughters Down syndrome, because then they would abort the child. For the Shenk's there was no point in risking miscarriage. Fact is his daughter was born healthy.

Already C.S. Lewis suggested "absolute biotechnological power as corruptive." Lewis described in his prescient 1944 essay, "The Abolition of Man," the final state is come when Man by eugenics, by prenatal conditioning has obtained full control over himself."

8 Legal twist in the increasingly complex struggle

8.1 Issues destructive for embryos and scientists

<u>Crime and Punishment</u> is the first great novel of the Russian writer and essayist Fyodor Mikhaylovich Dostoyevsky, published in 1866. It focuses on the moral dilemmas of the impoverished ex-student Rodion Romanovich Raskolnikov from St. Petersburg. The former student executes a plan to kill an unscrupulous pawnbroker for her money and argues that he freed the world of a worthless parasite. To counterbalance the crime Radion performs good deeds with the pawnbroker's money. Several times throughout the novel, Raskolnikov justifies his actions, believing that murder is permissible in pursuit of a higher purpose. Dostoyevsky (1821-1881) is acknowledged as one of the greatest psychologists in world literature.

Gotthold Ephraim Lessing (1729-1781) the first dramaturg (the art or technique of dramatic composition) was a German writer, philosopher, and publicist, and one of the most outstanding representatives of the "Enlightenment era" of the 18th century. He substantially influenced German literature. His ideological drama "<u>Nathan the Wise</u>" (Nathan der Weise), contains the Ring Parable, a fervent plea for religious tolerance. Everybody should reach eternal happiness by following his personal moral rules.

Sometimes tolerance seems to be tested, especially over the controversy of using human embryonic, versus adult stem cells in research. In this manner it cannot be ignored, that stem cell researchers react very sensitively to opponents and even attempt to silence them (Meeting of Society for Neurochemistry, 2005).

On August 25, 2010, two adult stem cell researchers Dr. James Sherely and Dr. Theresa Diesher, argued before the US government that the NIH guidelines broke the Dickey-Wicker amendment and that their careers were harmed by having to compete for government funding with researchers using embryonic stem cells.

Competition in San Diego, CA

District court judge Royce Lamberth, temporarily barred federal funding for studies on stem cells derived from human embryos that are later discarded. The judge ruled that it violates a law called Dickey-Wicker amendment, which prohibited the use of taxpayer money for experiments that destroy human embryos. With research grants on hold, the State University of Wisconsin banded together to fight for human embryonic stem cell research on September 7, 2010. Even a temporary suspension of federal funding for embryonic stem cell research has a far-reaching effect for the state argues Tim Kamp, a professor of medicine and director of the University's Stem Cell and Regenerative Medicine Center. Kamp said: "It penalizes taxpayers, who have invested so much in this research. And it penalizes patients who have been waiting so long for a treatment." The University of Wisconsin, Madison is lacking private funding and it would be a burden on researchers to stay in a part of a country that has less private funding for research, argues one scientist. "The UW-Madison has a strong tradition of academic freedom. Scientific inquiry coupled with ethical choices has to be allowed to continue," explained Martin, a pioneering stem cell scientist. Gov. Doyle reasoned that: "scientists deserve that funding's guide their research, political issues beyond their control are tremendously destructive. Even a short-term halt in federal support gives other countries with aggressive stem cell programs a shot," clarified the Governor, who further offers the state's help on any legal front in lifting the suspension. He underlines that he is talking about jobs for researchers and post docs who could be leaving for the UK or Israel. For the Governor the halt of tax-payers money to fund human embryonic stem cell research is a serious threat (Barncard 2010). And thus the Governor appointed on September 10, 2010 Attorney General J.B. Van Hollen to represent the state in its fight against the recent court ruling. But Van Hollen denied his help. His spokesman said the office had not received the materials "that would normally accompany this kind of requests," such as copies of pleadings, court filings, or a draft of the amicus brief the state would be joining. The General Attorney alleged: "It would be irresponsible for this office to step into a complicated, highly-charged dispute without having all the facts and information we require." Joe Heim a UW of La Cross political science professor is convinced that the Attorney General is just using an excuse. Dr. Heim said: "It sounds like the AG doesn't agree with the decision to appeal and is dragging his feet" (Spicuzza 2010). A day earlier on September 9, 2010, government lawyers fielded a request with the United States Circuit Court of Appeals for an "immediate administrative stay" of the ban. Scientists praised the ruling seemingly unaware that it is a temporary override of the injunction. The final decision might reverse or put the injunction back into place (which would once again stop any new embryonic stem cell research not already approved) (Kaiser 2010).

The hope remains that human embryonic stem cell research eventually could lead to cures. It seems that prohibiting the destruction of human embryos is harmful to scientists. NIH scientists were shocked reported the New York Times. The injunction appears to set back the scientific clock to President Bush's executive order restricting federally-funded research money to human embryonic stem cells. The Busch-era policy from August 2001 was overturned by President Obama's executive order in 2009. Obama did allow the National Institutes for Health (NIH) to set ethic guidelines over which cell lines would qualify for funding. And human embryonic stem cell research depends upon the destruction of human embryos (Adams 2010).

Ron Johnson the elected U.S. Senator in November 2010 stated: "My basic belief is you don't want to get in a situation where you are creating life through destroying it." Johnson supports adult and umbilical cord stem cell research. "If there is a program that's morally objectionable to a high percentage of the American public, that's probably something we shouldn't spend money on," Johnson declared (Ramde 2010).

8.2. At the forefront of scientific progress

It was quite embarrassing when Susan M. Reverby, a Wellesly College Professor of History and Women's Studies found out that US scientists infected 696 Guatemalans with syphilis in experiments conducted from 1946 to 1948. During that time, scientists sent prostitutes infected with syphilis into a Guatemalan prison, mental health hospitals and army barracks to test possible cures. Guatemalan law allowed "sexual visits" in such institutions. Dr. John Cutler conducted also the similar infamous "Tuskegee syphilis" medical experiments, where he observed 399 African Americans with late stage syphilis but didn't treat them. Even though the events occurred long ago, Secretary of State Hillary Clinton and Health and Human Service Secretary are outraged about the "appalling violations" of medical ethics. In 2010 U.S. government apologized to all the individuals who were affected by such abhorrent research practices (Johnson 2010).

The executive director of the Holocaust Museum in Houston, Texas, Susan Meyers reminds us in the last sentence of an article called "How Healing becomes Killing" that: "We must take care to consider the choices we make in coming years on issues of modern medical, ethical, scientific and public health policy in light of mankind's so obviously poor choices of the past."

The term "ethic of healing" was used by Nazi physicians working in concentration camps. In order to find a cure for soldiers with frostbite Nazi physicians performed experiments on prisoners. Most well-known philosophers and theologians never mention the now commonly used term "free decision of the conscience". They speak of the "dictate of conscience" which one should be compelled to follow in doing the right thing by applying ethical standards of good and evil to specific situations. Our judgment however needs to be based on an informed conscience. We thus have the duty to educate and support our conscience and to follow its instructions (Spaemann 2008). Some physicians working in the concentration camps quit their job, as they felt obligated to their conscience in responsibility to their professional ethos. Even so the Nazi ideology defined their duties and professional obligations, which however were contrary to those physicians' consciences (Spaemann 1999).

Max Weber is one of the most profoundly influential thinkers of the twentieth century. Born in Germany, Weber became a lawyer, politician, scholar, political economist, and sociologist. He spoke nineteen foreign languages and became a major scholar of religion as well. He suggested two sets of ethical virtues, the ethics of responsibility and the ethics of conviction. The ethics of responsibility deals with a possible causal effect of an action. Elements of our action can be reoriented in order to achieve the desired consequences. An ethical question is thereby reduced to a question of technically correct procedure, and free action consists of choosing the correct means.

In the ethics of conviction the action is bound to the doctrines of morality and values. The freely acting person is concentrating on the means but not on the end. Max Weber labeled the ethic of conviction depending on the values of the historical context the person is living in. Thus, it is changeable and Max Weber compared the set of ethical convictions with an air sleeve (Max Weber 1919).

In case a researcher wants to order human embryonic stem cells and is searching the pertinent online sources, he will find on the webpage various links to statistics and surveys. In conjunction with the services, he will automatically get the information how many American Catholics promote human embryonic stem cell research, despite the veto of the American

Bishops' Conference. Although it is mentioned that more Protestants are in favor of it. Scientists deny their opponents to speak out against stem cell research and question the right of Catholic Bishops to speak for all Catholics including Catholic scientists. It is further stated, that: "from the earliest days of the Church, Catholic theologians have been at the forefront of scientific progress". Listed are: Tertullian (160-225), Saint Augustine (354-430), Thomas of Aquinas (1225-1274). Whereas Saint Hildegard from Bingen (1098-1179) or the famous Augustinian monk from Brünn, George Mendel (1823-1844), the founder of biogenetics, are not mentioned. Many other Catholic scientists who developed new and exciting discoveries that enrich our life's today followed them. It shows that Catholics were at the forefront of the scientific development. Even the Catholic Church is encouraging scientific discoveries. Under this aspect, researchers cannot understand the immense controversies about human embryonic stem cell research. For them the Church should not be opposed to sciences (O'Brien 2008).

The future will show whether we can reach the goal of employing human embryonic stem cells as therapeutics, or if we were irresponsible and unfaithful with the human gene-pool entrusted to us. Currently we have only therapeutic success in working with adult stem cells. Future generations will either praise the researcher or their opponents and will judge and condemn either of them. Thus, maybe the point of view of the acceptability of embryonic stem cell research will change and in hindsight, we will respect and be grateful for the moral teaching of the Catholic Church and its opposition to embryonic stem cell research.

But even today, very famous Catholic scientists uphold the bishop's opinion. The renowned British Catholic stem cell scientist Colin McGuckin, professor of regenerative medicine at Newcastle University, recently moved to France. With this move, he protested against the U.K. Universities and funding agencies, which continually prioritized embryonic stem cell research over his work on adult stem cells. For him the University acts independently despite the superior clinical success of adult stem cells. According to him, adult stem cells are, up until now, the only cells that are showing therapeutic success. He concluded that they are not recognized at all and the research is totally underfunded. In 2005, McGuckin isolated for the first time fetal stem cells from the umbilical cord blood. In 2009, the Professor and ten of his colleagues started at a new research center in Lyon, France in a much better environment. He remains very critical of the one-sided funding in England. For him the priority is to heal patients with adult stem cells, which have been shown to be effective. That goal he could not reach. It was even impossible to get support for his scientific work in the exemplary stem cell research country Great Britain. The scientist is furthermore convinced that it is not necessary to destroy embryos for the purpose of healing diseases (Waggoner 2008).

The University of Wisconsin (UW) in Madison, USA has a major stake in embryonic stem cell research. Professor James Thomson isolated the first cells and he remains a leading international authority in that field. More than 100 scientists are working in thirty-five to forty research groups on that topic at the University campus. The UW WICell institute is a nonprofit research institute that supports the study of stem cells. They received more than $38 million in federal funds for investigations since they were approved for federal funding. WICell was the National Stem Cell Bank with eighteen of twenty-one NIH supported embryonic stem cell lines. The bank distributed embryonic stem cells to other investigators. It also provided technical assistance to researchers. "In 2007 additional $44.5 million research funds were provided," stated

Andy Cohn, president of the Wisconsin Alumni Research Foundation. "Eventually this research is expect to lead to breakthroughs in the treatment of Alzheimer's disease. More federal money is promised for the campus of UW Madison and so one of the last obstacles is taken to reach the goal to being successful" (Pitsch 2008). It is interesting that of all the diseases only Alzheimer was mentioned in the Newspaper article from 2008. Researchers from the University of Alabama, Birmingham, from the Research Institutes of Jülich and from the University of Düsseldorf in Germany, collaborate to find therapies for Alzheimer. Currently Alzheimer is incurable. Only the symptoms of the disease can be alleviated. The causes of Alzheimer's are plaques called abeta (amyloid beta) deposited throughout the brain of the patients. The scientists developed a peptide, consisting of D-enantiomers amino acids (D3), which are capable of binding and destroying abeta. Also inflammatory processes are declining under the peptide "D3". The experiment has to be converted from mouse to human (Willhardt 2008).

8.3 Oocyte Sharing

Ovaries are the limiting resource of human embryonic stem cell research. It is very difficult to obtain them. They are important for the progress in basic science research. In Great Britain and in the United States egg cells for research purposes are donated from in vitro fertilization clinics. The Roslin Institute in Scotland, made famous by Dr. Jan Wilmut and his Dolly-method for cloning, is attached to an in vitro fertilization clinic. It is one of the international leading stem cell institutes, also collaborating with research centers in the United States. Dr. Wilmut, by training an agricultural scientist, regrets the lack of collaboration within Europe, largely because of legal differences regarding stem cell research. According to him, the situation is very debilitating. Imminent success will be only possible when all states work together (oral report by Dr. Wilmut). The statement sounds similar to argumentations of Soviet Communism: "The creation of a worldwide communist paradise can only be reached if all countries and people of this world become united in communism."

The institute of Dr. Wilmut was established to gain a fundamental understanding of the genetics of cellular and organ bioscience. The goal of the research center is to increase the health and welfare of farm animals with bioengineering. Comparative biology is focusing on the basic mechanisms in order to gain knowledge of the genetics responsible factors for resistances and diseases in animals and humans. They found that breeding would increase when animals are resistant to parasites, and fed with a balanced nutrition. They also studied zoonotic diseases (pathogens crossing from animals to humans). They further showed that improvement of animal health in a holistic way is beneficial for the environment, farmers and agriculture. An increase in the food production is profitable for humans themselves. The Roslin Institute was founded in 1993, independently from the University of Edinburgh.

Genetic engineering is embraced by only a few European and African communities and in fact violently opposed by many. These new technologies require regulatory policies called bio-politics, which aim at the proper use of genetically manipulated organisms. Scientists and the rural populations increasingly consider the political significance of recent advances in biological sciences. However, emerging bio-political insecurities started the dialogue between science and the public to dispute the consequences and to explore alternatives. European countries, for example, delayed the approval for planting bioengineered crops. Swiss consumers reject genetically manipulated food. In contrast, in November, 2004, a law was immediately dismissed which allowed the destruction of seven-day old human embryos. With the highest possible agreement, the Swiss National Council accepted this law. Twenty-three parliamentarians voted for the new law, zero against and twenty-two parliamentarians were not present to vote. There were no advising ethic commissions or even a passionate debate. Mass media sparsely reported the consequences of changing a law influential for the future of the human genome.

Ethical commissions are rather concerned that women donating their eggs do not get reimbursed. The demand of eggs is growing. Women either donate eggs for stem cell research or for infertility centers. The number of donors is relatively low. The egg extraction process is invasive and carries certain risks. Women normally produce a single mature egg per cycle. In order to maximize success rate, several ovarian stimulation medications are used to increase the ovaries to make sufficient follicles. Reproductive facilities prefer to get 8-15 "high-quality eggs". Fertility doctors need to control the women's hormone levels and reproductive system's functions in order to create a hospitable environment for ovarian stimulation. They administer gonadotropins to produce a larger number of eggs in the donor's ovaries. The drug used, a synthetic copy of the GnRH found in pig and sheep, is a nonapeptide analog with the sequence: p-Pro-His-Trp-Ser-Tyr-D-Leu-Leu-Arg-Pro-NHEt. It is synthetically modeled after the natural hypothalamic gonadotropin-releasing hormone (GnRH). GnRH interacts with its specific receptors to first suppress the pituitary gland to release hormones FSH and LH. This step is important for the recruitment of multiple follicles for a proper maturation. Because decreasing the body's production of specific hormones and natural chemicals will influence the behavior of certain cells. Synthetic GnRH is commonly used to shut down a woman's ovaries before stimulating them to produce multiple follicles. Complications can and do occur, because this drug causes a range of problems including: rashes, vasodilatation (enlargement of blood vessels causing a "hot flash"), burning sensations, tingling, itching, headaches and migraines, dizziness, hives, hair loss, severe non-inflammatory joint pain, difficulty breathing, chest pain, nausea, depression, emotional instability, loss of libido, dimness of vision, fainting, weakness, asthenia gravis hypophyseogenea (severe weakness due to loss of pituitary function), amnesia, hypertension, rapid heart rate, muscular pain, bone pain, abdominal pain, insomnia, swelling of hands, general edema, chronic enlargement of the thyroid, liver function abnormality, anxiety, and vertigo (Norsigian 2005, a). In vitro fertilization does not work when only one egg is harvested. The extracting of multiple eggs however is very risky, but it increases the likelihood of success. Also for research purposes, it is important to have a large number of eggs.

We are at the early stages of embryonic stem cell research with only hypothetical benefits at hand. It might be even wiser only to extract a single egg each month. Ethics committees have to oversee the risk/benefit ratio. Young women donating multiple eggs might end up with tragic consequences that even affect their fertility. Is it justified to take that risk of infertility? Advocates believe that fully informed consent is not possible, even not ethical because we have not enough data on gonadotropins. "This drug was developed for women to treat symptoms of endometriosis (overgrowth of uterine lining outside of the uterus) or uterine fibroids and has no current FDA indication to be used in fertility clinics for super-ovulation" (Norsigian 2005, b). The greatest advantage of GnRH-based technology is in the use of long-term fertility control in female animals. Lupron has a broad range of applications in production and domestic animal management (Herbert and Trigg 2005). "Women undergoing the Lupron treatment venture on a dangerous path. It is a concern that some potential donors are not fully informed because of the fear that fewer might volunteer," stated Judy Norsigian, the Executive Director of Our Bodies Ourselves, a women health education and advocacy organization.

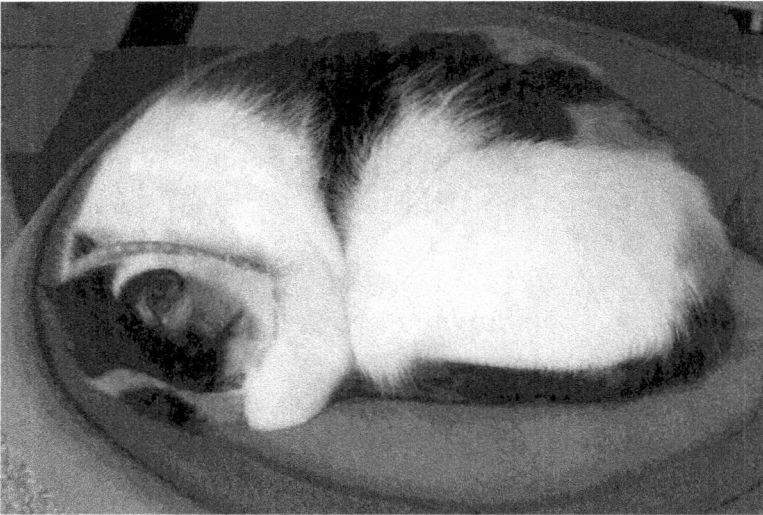

Ignored side effects

Some consider it wrong to pay egg donors, others consider it fair. Without willing donors, there will be less research on human embryonic stem cells. Oocytes are usually donated for reproductive purposes (oocyte sharing). The donors will be recompensed with at least $5,000 per cycle; or will receive a discount for personal in vitro fertilization. The notoriety of the Korean scientist Dr. Hwang raised public awareness of egg donation for research (Magnus 2005). The public considered Dr. Hwang's use of oocytes from his dependent research assistants an ethical offense. The success of Dr. Hwang, Professor of Veterinary Medicine at the University of South

Korea, was due to the use of particularly fresh, unfrozen oocytes.

In the United States, oocytes and embryos are typically culled from fertility clinic leftovers otherwise destined to be thrown away. In 2005 the National Academies of Biomedical Ethics recommended in their Guidelines for Human Embryonic Stem Cell Research that no payments should be provided for donating eggs, sperm, or embryos for research. Jonathan Moreno, co-chair of the committee and Professor in ethics at the University of Virginia, is convinced that his recommendations are justified by the sensitivity of egg donation for stem-cell research. In addition, there is a huge uncertainty about the actual risk of severe complications donors have to deal with (Steinbrook 2005). The compensation for egg donation is an ongoing contentious issue, because sperm donation and other research subjects are usually compensated. Bioethics argue that egg donors should be compensated as they endure a lot of discomfort. In 2004 reimbursement for egg donation was banned in South Korea, Massachusetts and California. Some researchers see the passions and pressures associated with stem-cell research more profoundly. This results in secretive payments and black market practices. Bioethics Professor Bonnie Steinbock from the State University of New York/Albany is opposed to such practices (Steinbock 2004). In her opinion there is no child coming from the donation, so nobody benefits directly from the donor. However, in 2007 it was decided that women are to be paid to donate their eggs for scientific research. Even so, scientists from the University of Padua, Italy have warned that women who donate their eggs for research could be at risk from life-threatening side effects induced by the powerful drugs that may even be fatal. In addition, poor women might be tempted or coerced to participate. They do it in the hope of earning easy money and thus a new kind of an oocyte exploitation of third world countries is created (Campell 2007).

There is an interest in using eggs for somatic cell nuclear transfer. The donated oocyte nucleus will be removed and replaced by a diploid nucleus from the patient to create embryonic stem-cell lines. The cloned embryo has patient specific attributes. The procedure is known as therapeutic cloning. Because the embryo is going to be destroyed within fourteen days, it does not matter if the egg is coming from India, China or Africa. It is even thought to create egg cells from human embryonic stem cells. However, this research is in the preliminary stages. Another big concern might be that donating females are primarily the friends or relatives of persons with diseases or disabilities. The promise of scientific breakthrough in embryonic research motivates them. Thus, it may become beneficial for women with genetic diseases to donate their own oocytes (Pennings 2007). However, only rare autosomal dominant diseases like night blindness and brachydactyly (short hand syndrome) can benefit from egg donations, because only autosomal dominant diseases are fully manifest in the egg cell and thus can be properly studied. It is unlikely that donated oocytes should be utilized for such rare conditions. An autosomal recessive hereditary disease is for example cystic fibrosis (mucoviscidosis). In order to study autosomal recessive diseases, sperm and egg cells have to be carriers.

8.4 Intermezzo bioethics-science

An advantage for adult stem cells is that they are taken from the patient and thus will not be rejected. Even so, scientists believe in the pros of human embryonic stem cells because they might morph into any cell of the body. Scientists hope to harness them so they can create replacement tissues to treat a variety of diseases. Scientists are facing similar difficulties with organ-transplants. Many people are waiting for a life-saving transplant. Organ rejection has to be prevented by immunosuppressant medication. It is feared that stem cell therapy when finally ready to be used will influence the host immune system. A very challenging task for embryonic stem cell researchers is to circumvent the immune system. Cloned embryos and embryonic stem cells are completely foreign cells. They have their own immune system, like each egg and sperm cell. Hypothetically, embryonic stem cells can only be transplanted with an additional immune suppressing medication. Medications have a long list of side effects. Researchers tried to bypass the immune system by the technique of therapeutic cloning. Therapeutic cloning is a technique to create cloned embryos, genetically identical to the patient and thus the isolated human embryonic stem cells are tailored to a specific patient. The patient would only have to donate his own somatic cell nucleus. The DNA with the disease causing genes needs to be repaired prior to cloning. The patient DNA is then inserted into an egg whose nucleus has been removed. Now the egg cell has a diploid nucleus like all embryos. For the cell line itself, more than one nuclear transfer is required.

Another challenge is the incompatibility of the mitochondrial DNA. Mitochondria are cell organelles in the cytoplasm. They possess an inner and outer membrane and even their own DNA. The mitochondrial DNA is inherited exclusively from the mother. Mitochondrial DNA has been called a biological history book of women. In 1987 three California biochemists proposed a new way of tracing back human origins by using mitochondrial DNA. Fortunately, occasional mutations occurred otherwise the world would have identical mitochondrial DNA. By tracing all the families back, one could arrive at the original ancestor. Analyses led back to a single ancestral woman living in Africa about 200,000 years ago. Biochemists proposed a new way of tracing human origins in the so-called mitochondrial Eve theory (Cann et al. 1987). Each mutation establishes a new mitochondrial family.

In July, 2010, Harvard Medical School described that mitochondrial DNA is unique among endogenous molecules because mitochondria evolved from prokaryotic ancestors. The similarity of bacterial DNA to mitochondrial DNA bear bacterial molecular markers. Like bacterial DNA, circulating mitochondrial DNA can activate an innate-immune stimulatory "danger" response. Dr. Zhang hypothesized that shock-injured tissue, a traumatic or hemorrhagic shock cause the release of mitochondrial DNA. When circulating in the body it contributes to the initiation of a systemic inflammatory response syndrome. (Zhang et al. 2010). In March, 2010, Dr. Zahng explained to Nature Journal that cellular injury like a disruption releases "mitochondrial enemies" within (he means the mitochondrial DNA) that activate immunity and causes the immune system to respond in a way much like sepsis. Associated with organ failure it is the leading cause of premature death (Zhang et al. 2010)

For the first time it was described that mitochondrial DNA provokes an immune response. Initially, it will build up a line of defense that causes a systemic inflammatory response. The presence of hundreds of copies of mitochondrial DNA in each human cell is quite a challenge. In 2010 researchers discovered a widespread heterogeneity in the mitochondrial DNA. Contrary to normal tissue, cancer-derived tissue harbored mutations, which demonstrated that a complex mixture of related mitochondrial genotypes characterizes individual humans. The study provided insights into the nature and variability of mitochondrial DNA sequences that will implicate mitochondrial processes during embryogenesis, cancer, biomarker development and forensic analysis (He et al. 2010). Nobody knows how mitochondrial DNA, still present in the enucleated (animal) egg cell will react to a foreign cell nucleus containing the DNA of a human being. Dr. Hwang, the Korean clone expert denied the interference of mitochondrial DNA from the enucleated egg cell with the foreign cell nucleus when he was asked after a talk he gave at a scientific Neuroscience Meeting in Wisconsin in 2005. Therapeutic cloning is forbidden in Germany. A violation of that law will be punished with a term of imprisonment for at least five years, or a monetary penalty (Hornbergs-Schwetzel 2008).

On November 15, 2008, Great Britain passed a new Human Fertilization and Embryology law. It allows: "animal-human embryos, artificial gametes, cloning using two maternal egg sources, germ line manipulation, pre-implantation diagnosis for eugenic purposes, posthumous conception, removal of the child's need for a father, use of tissue without proper consent" ..etc. "Because human egg cells are difficult to obtain it seems much easier to harvest animal eggs from cow ovaries at the local abattoir. They are available in much greater numbers". These egg cells are used to create cytoplasmic admixed embryos (animal-human embryos). However, not only cow eggs are used. In China rabbit eggs are utilized (Wilmut 2007, personal communication). In cytoplasmic admixed embryos, the animal nucleus is removed and replaced by a human somatic diploid nucleus. Parkinson or Alzheimer patients are eager to provide their somatic cells for this procedure. Human-animal embryos possess the mitochondria from the animal. The purpose of these techniques is to overcome the shortage of human egg cells. It is curious, that the influence from the mitochondrial DNA inherited from the animal cell is disregarded in this new technique.

For a long time the method of nuclear transfer worked only in mouse models. Researchers from the Howard Hughes Medical Institute in Boston see a need and want be able to clone patient specific human embryos. In 2004 South Korean clone scientist Dr. Hwang Woo-Suk claimed that he had generated the first human embryonic stem cells from healthy people, and in the following year from afflicted patients themselves, using more than 100 nuclear transfer trails in his abbreviated cloning method. The news was huge, but the information was simply dishonest. Later one it was clear that his research results were factored. In 2007, the Oregon Health and Science University stated that after years of unsuccessful attempts they were able to generate monkey embryonic stem cell lines genetically identical to the donated nuclei from another adult rhesus monkey (Baker 2007).

There are two main purposes for cloning: medical therapy and reproduction. Both are using "somatic cell nuclear transfer" (SCNT) to create genetically identical animals. When the sheep Dolly was cloned the immediate reaction was that humans are now able to clone themselves. Even if it would be considered ethical, reproductive cloning is simply too risky, because of high rates of deformities, disabilities and death observed in animal cloning. Identical twins are an example of human clones that are created naturally. Dolly was a symbol of having accomplished great possibilities in science. It opened the door to a bright future in the field of biotechnologies in creating something like replacement parts. It is believed that by cloning techniques the pinnacle of stem cell research is reached even through human reproductive cloning is considered universally unacceptable. Since 1991 Germany emphasizes that it is illegal to clone humans, in order to protect human embryos. In vitro fertilization allows the only creation of embryos that are actually used. The import of human embryonic stem cells is possible, but under strict surveillance. In April, 2008, Germany decided that stem cell lines can be legally imported when created until of May 1, 2007.

Britain, Canada, New Zealand, several US states, Singapore and South Africa, make a distinction between "therapeutic" and "reproductive" cloning. Human clones, created in a lab must be killed within a certain period of time (usually fourteen days), and cannot be implanted in a woman's womb. That means created human lives are destroyed and not given a chance to live. The technical procedure of cloning is the same for "reproductive cloning" and "therapeutic cloning". It seems incomprehensible to support therapeutic cloning but to oppose reproductive cloning. Furthermore there is no logical reason to prohibit cloning if we disagree only in the purpose of it said Clarence P. Jeffrey a writer for CNS (White 2009).

Dr. Severino Antinori Buero, a gynecologist from Italy boasted that he was able to clone a human being, who was born in January, 2001. He cloned a baby in the same way as you would clone an animal. He insists having a success rate of ten percent. Davor Solter a scientist from the German Max-Planck-Society in Freiburg, stated: "It is ethically untenable to repeat animal experiments on humans. Such trials are accompanied by a high failure rate". Jan Wilmut the father of "Dolly" claimed his success rate with 0.36%. He needed 277 embryos to produce the cloned animal.

Dr. Antinori is convinced he did not commit a crime in cloning humans. In addition, he refuses to give a statement whether the cloned children are healthy. He declared that it is not within his power to give any information, because he is lacking long-term results. In addition, he refused to reveal the location of the cloned children in order to protect them from the public. Antinori Buero views it as a human right to have children. His research is dedicated to overcoming infertility.

He refers to a Chinese colleague, the female scientist Guang Zhou, who allegedly cloned 400 embryos. It seems that the Professor did not notice that Guang Zhou is the city Canton in South-China. The only truth in his statements is the fact that he helped a sixty-three-year old woman to become pregnant. It was the first postmenopausal pregnancy in history. Dr. Buero still asserts that over 100 children were cloned by his method. The existence of the cloned children is doubtful and could never be verified. This incident resembles recent news from an obscure sect, which claimed to have cloned a baby named Eve (Ladstätter 2003).

Craig Venter a famous US genetic researcher declines to appraise the challenges to clone humans. He said, "to start cloning experiment on humans is not vindicated because its social justification is not given". Venter, the pioneer in decoding the human genome, is president of Celera Genomic Cooperation in Rockville, Maryland. He demands that the US government pass a law as soon as possible that protects human's hereditary factors against discrimination. People may lose their health care if their genetic risks are known (Berner Zeitung 2001). Venter is not the only one with that opinion. Representatives of the green-democratic party from Regensburg, Germany, vehemently attack cloning attempts. For them it is an act against humanity. When in 2005 Jan Wilmut received the taxpayers funded 100,000 Euro, Paul-Ehrlich-Award from the Frankfurt University/ Germany, environmentalists demonstrated against it.

9 Hypothetical and highly experimental

9.1. Researchers claim therapies for the future

In 1998, the first successful isolation of human embryonic stem cells initiated one of the most controversial biotechnologies of our generation. The new technology appeared to have many promises according to some scientists. These cells were considered to possess the capacity to morph into many different types or tissues of the human body. It is speculated that the cells offer possibilities to produce new lifesaving therapies. The plethora of diseases and medical research problems might potentially have benefits from this research. The gains from stem cell research are unlimited and could be used to replace degenerated cells and tissues in the human body. If scientists were only able to solve the obstacle that large-scale implementation of stem cell therapy to a clinical setting require: the establishment of a reliable and controllable method of differentiating cells. For this reason no therapies, however, have been developed. Critics of the research say that destroying embryos to advance academic study is immoral. Alternative approaches, such as using stem cells derived from adults, were equally or even more promising. Human embryonic stem cell researchers believe that the only handicap for not being as successful as previously supposed is due to the lack of appropriate funding. "We were really looking forward to the next chapter when human embryonic stem cells could really be explored for their full potential," said May Comstock Rick, immediate past president of the Coalition for the Advancement of Medical Research, a group that has been lobbying for more federal funding (Stein and Hsu, 2010).

Meanwhile human embryonic stem cell researchers try to make human embryonic stem cells secure, because only a safe and efficacious therapy will save human life, as they state. "The humanitarian spirit will apply the best of mankind's technology in the alleviation of human suffering," said Butler, the president of the international Longevity Center in New York and cofounder of the National Institute on Aging (Butler 2008). Is it possible to reach that goal, by destroying human embryos? Adult stem cell researchers are very suspicious. The best that we have is the future generation. Why would researchers deprive embryos of their right to life in order to keep old people alive longer? It has to be considered that destroying one life for the benefit of another violates the conscience or moral religious beliefs of some researchers. The "conscience protection" has raised questions whether doctors, pharmacists and other health care workers have the right to refuse to provide services that conflict with their religious beliefs. The tension to perform an act that is perceived as life-destroying incidentally did burden health care professionals during the Nazi time. Can society force researchers to act against their conscience? It is thought provoking that many people in the Third Reich did suffer so much because they acted against their conscience. The sanctioned moral standards were imposed on them by the regime, which limited their freedom for a different viewpoint regarding morality and ethics.

In 1961, President Kennedy had a vision of the first man landing on the moon within a decade. For that goal, the president generated his financial resources. Can we compare stem cell research with the trip to the moon? There is the risk that the costs may not justify the benefits. Also the question remains whether we have the permission to destroy nature, in this case the gene pool, without the proper knowledge of its true benefit.

**Geological profile of Grand Canyon
provides insights of our Earth**

Besides, the cost benefits of the moon missions were very high. Was it undertaken to prove that humankind should be allowed to do everything technology enables them to do?

The biggest pay off so far was with human embryonic stem cells. We know that embryonic stem cell research is very expensive and is even impaired by moral objections by a majority of Americans. In the US no federal law directly regulates human embryonic stem cell research. The federal policy is based on the so-called Dickey-Wicker Amendment. In 1999 the Amendment clearly prohibited the use of federal funds for all research in which a human embryo is destroyed. Nevertheless, policy on misguided legal interpretation spoke about a "responsible, scientifically worthy human stem cell research, including human embryonic stem cell research, to the extent that it was permitted to be funded" (Ogbogu 2010). In 2009 Prof. Geoffrey Raisman, director of the Spinal Repair Unit at the University College London Institute of Neurology stated: "The attraction of human embryonic stem cells lies less in their therapeutics than in their imagined commercial potential - but is this justified by science?"

9.2 Technical hurdles to the promises of human embryonic stem cells

Embryonic stem cells are derived from the inner cell mass (embryoblast) of the five-day-old embryo known as a blastocyst. The blastocyst includes three structures: the trophoblast, which is the layer of cells that surrounds the blastocoel, which is the hollow cavity inside the blastocyst; and the inner cell mass. This blastocyst has to be destroyed for the isolation of stem cells. With that also the cellular support system, the trophoblast disappears. The trophoblast usually forms the placenta (Breburda et al. 2008, 2006,b), which supports the embryonic and fetal development (Breburda et al. 2009). The blastocyst consists of 50-150 cells. A group of approximately thirty cells generates the embryoblast. Embryonic stem cells are pluripotent, because cells obtained from the embryoblast are able to differentiate into all derivatives of the three primary germ layers: ectoderm, endoderm, and mesoderm. Pluripotent stem cells have the potential to differentiate into more than 220 kinds of human body cells. Contrary to human embryonic stem cells, adult stem cells are multipotent progenitor cells. These cells can display a number of cell types with limited differentiation. They are present in every adult organism.

When no stimulus of differentiation is added to embryonic stem cells, they remain pluripotent through multiple cell divisions. Because of their plasticity and potentially unlimited capacity of self-renewal, stem cell research is proposed to be the appropriate therapeutic remedy. To date however, no approved medical treatments have been derived from embryonic stem cells. In 2008, the president of the International Society of Genetics mentioned at the meeting in Berlin that he is hopeful that embryonic stem cells are going to be the therapy of the future. For instance, dopamine producing embryonic stem cells could be used to treat patients with Parkinson's. Others view this therapy to be ready in only two decades.

Adult stem cells and cord blood stem cells have thus far been the only stem cells successfully treating many diseases. Still many people believe that embryonic stem cells offer the greatest promise for the developing of new medical therapies. While actually adult alternative sources of stem cells have demonstrated a much brighter prospect. Lord Alton stated in 2009 that there are currently over eighty therapies and around 300 clinical therapies under way using adult cells. "We still have no stem cells at all from cloned human embryos, never mind routine clinical treatments based on cloning. This sharply contrasts with burgeoning number of treatments using adult stem cells."

The growth of cells in the laboratory is identified as cell culture. Isolated human embryonic stem cells are transferred into a plastic laboratory culture dish containing nutrients. In the culture medium, cells proliferate and spread over the surface of the dish. Differentiation is not taking place, because scientists are still more or less puzzled over the right agents to accomplish that task. Stem cell growth media has to fulfill certain requirements in order to maintain pluripotency of the cells. Historically, typical feeder layers include mouse embryonic fibroblasts or human embryonic fibroblasts. A fibroblast is a kind of cell that produces extracellular matrix and collagen and plays a role in wound healing. The inner surface of the culture dish is coated with mouse embryonic skin cells, which have been treated in order not to divide. These cells have a sticky surface. The procedure is bearing the risk that retroviruses from the genetic makeup of the mouse might be able to infect human embryonic cells. Thus, scientists have begun to devise ways of growing embryonic stem cells without the mouse feeder cells.

In 2010 Prof. Kiessling from the University of Wisconsin, Madison reported about a new culture system utilizing a synthetic, chemically made substrate of protein fragments, peptides, which have an affinity for binding with stem cells (Devitt 2010).

After the cells grow in the culture dish they are gently removed and placed into several fresh culture dishes. Repeating that procedure for each subculture cycle will yield millions of embryonic stem cells. After six or more months, the pluripotent stem cells generate embryonic stem cell lines. Once established, they can be frozen or shipped to other laboratories for experimentation. A stem cell line costs $5000 to $6000. It is known that human embryonic stem cell lines are difficult to grow in culture. The cells surviving the thawing process from liquid nitrogen is only 0.1-1% (Karberg 2008).

These cells are called immortal, because they have the capacity of long-term self-renewal. As long as the cells are grown under these conditions, they remain undifferentiated. They are able to differentiate spontaneously when they are allowed to clump together. So-called embryonic bodies are established. Another possibility is to inject undifferentiated (human) embryonic stem cells into a living vector (mice) in the hope the cells will adapt the structure of the surrounding cells. The immune system of the lab-rodent is switched off to prevent rejection of the cells. Embryonic stem cells transform into so-called teratomas. Teratomas contain a mixture of many differentiated or partly differentiated cell types from more than one germ layer. The procedure was mainly applied to show the differentiation potential embryonic stem cells have. However, it is proven that undifferentiated embryonic stem cells, once injected into an organism, will form malignant tumors, because their differentiation is undirected. Undifferentiated cells are not applicable as therapeutics. Organized differentiation and proper embryonic development only occurs intrauterine. A chaotic differentiation in rodents will build a teratoma.

Currently the major goal for embryonic stem cell research is to achieve and control directed differentiation of human embryonic stem cells into specific kinds of cells. This is a difficult task that must be met if the cells are to be used as therapeutic remedy. Otherwise any therapy based on the use of human embryonic stem cells would only become hypothetical and highly experimental. The techniques being tested are borrowed from in vitro trials to direct the differentiation of mouse embryonic cells. Mice are considered to be the most important model organism for embryonic stem cell differentiation. Unfortunately, it seems to be more than true that it is a long way from mouse to man. At a fundamental level mice are similar to Homo Sapiens. Ninety-nine per cent of the mouse genes are matched by a corresponding sequence in human genome. This means that scientists are only allowed to conduct research on human embryo stem cells after they have clarified their specific question by using animal cells as far as possible. However in 2010, Hans Schöler at the Max Planck Institute for Molecular Biomedicine puzzled over to what extent the findings of studies on embryonic stem cells of mice are transferable to humans. They have both an active Oct 4 transcription factor, the gene that is essential for maintaining pluripotency and makes them potentially immortal. But certain signaling substances that can be used to turn mouse cells into more differentiated cells have either no or completely different effects in human embryonic stem cells (Gerber et al. 2010). In 2007 a research team isolated a new type of pluripotent mouse cells, known as epiblast stem cells. Only they are not from a few-days-old embryo in a blastocyst stage, they are harvested from an embryo already implanted in the uterus (Brons et al. 2007). Epiblast cells are a step further in differentiation and the lead author of the study Boris Greber assumes that his new cells are more-or-less equated with human embryonic stem cells. To prove the equality, Gerber, a biochemist, and his team tested how mouse epiblast and human embryonic stem cells react to different growth factors and inhibitors. They found that FGF (fibroblast growth factor) actively supports the self-renewal of human embryonic stem cells, whereas this was not the case with mouse epiblast cells. Ultimately that means that many preliminary tests on animal cells in medically relevant projects are misleading, says Hans Schöler. For him this kind of experiment showed that human embryonic stem cell research will continue to be absolutely essential for stem cell research in the future. "The recent success in reprogramming mature human somatic cells sometimes makes it look as though tests using human embryonic stem cells are nowadays redundant. But appearances are deceptive. Our latest study demonstrated that animal model systems are inadequate for a great many tests of that kind. Particularly when we are talking about developing safe and effective stem cell therapies, we still need human embryonic stem cells as the gold standard against which to compare everything else. In such cases, lengthy preliminary testing on animal cells risks wasting valuable time and resources," concluded the German scientist Hans Schöler (Max Planck-Gesellschaft 2010).

Researchers permanently assured the unique character of human embryonic stem cells. A great deal of time has passed since they were first derived and many questions concerning the nature of these cells persist. Through years of experimentation scientists have tried to obtain directed differentiation. Researchers changed the chemical composition of the culture media to stimulate differentiation. There are many technical hurdles between the promise of stem cells and the realization of their uses. To this day, the primary goal is to identify how undifferentiated stem cells become differentiated. Highly controlled modifications in gene expression are not known. Control of differentiation is an important area of current research. The goal is to understand the role and regulation of gene expression that channel regulatory signals from activator and repressor proteins have in effecting changes in gene expression programs that control diverse physiological processes, including cell growth and homeostasis, development and differentiation. And all of that is last but not least epigenetically restricted (influenced by the environment not the DNA).

9.3 Monopoly of embryonic stem cell lines

In 1981, the first successful isolation of murine embryonic stem cells took place. Another seventeen years passed before the first human embryonic stem cells were isolated. In 1998, Drs. John Gearhart from Johns Hopkins University and James Thomson from the University of Wisconsin first derived human embryonic stem cell lines. The scientific evaluation and close examination of the pluripotent cells were accompanied by many contentious ethical debates. Dr. Thomson and coworkers derived stem cell lines from embryos donated by couples undergoing in vitro fertilization. Dr. Gearhart and his coworkers derived pluripotent stem cells from fetal gonadal tissue destined to form germ cells. The tissue was obtained from aborted children. Pluripotent cells can develop into the three major tissue types: endoderm (which goes on to form the lining of the gut), mesoderm (which gives rise to muscle, bone and blood) and ectoderm (which gives rise to epidermal tissues and the nervous system). Embryonic stem cells are precursors to any one of many different cell types such as muscle cells, skin cells, nerve cells, liver cells. Researchers are still at the beginning of trying to understand how primitive stem cells have to be stimulated to become specialized. Scientists see almost infinite value in the use of embryonic stem cells to understand human development and the growth and treatment of diseases. In 2008, Dr. Thomson requested an embryonic stem-cell patent at the European Patent Office. He fielded his petition by a University of Wisconsin-Madison affiliate, the Wisconsin Alumni Research Foundation (WARF), which helps commercialize the University's research. The board of appeals for the European patent regulators rejected the patent request, arguing: "Embryonic stem cells are thought capable of morphing into any part of the body. While they are considered medically promising, embryonic stem cells are controversial because embryos are destroyed to harvest them." The European Patent Convention (EPC) does not allow patenting inventions whose commercial exploitation would be contrary to public order ("ordre public") or morality. Furthermore, the Convention prohibits patenting on uses of human embryos for industrial or commercial purposes (European Patent Office 2008). The WARF statement does not mention the concern of the patent office regarding the disturbance of the public order and morality. Furthermore, WARF claimed no existing research cooperation with Europe. WARF declared that the decision of the European Patent Office would not affect in any way WARF patent rights in the United States. The European Union patent rules are peculiar to Europe says WARF. WARF General Counsel Michael Falk stated that they have more than forty issued patents directed to embryonic stem cells in twelve countries with more than

200 cases pending all over the world. The European rejection would only represent a ruling in just one of these cases and in just one jurisdiction on grounds that do not apply in other jurisdictions (WARF 2008).

Adult stem cells are licensed for stem cell patentability, because no human life is destroyed in the process of their creation. The European "Green-peace" is very pleased with the decision of the EPC. They speak of a huge success and a milestone for patent rights. The Board of Appeals of the European Patent Office has underscored the ethical values in patent law.

The Wisconsin Alumni Research Foundation entered a licensing agreement with biotech companies and other researchers to commercialize stem cell technology. The contract allows using licensed patent technologies developed by Dr. Thomson. Fees are paid for products using the intellectual property and technology laws. The money is used to fund more research at the University. WARF has two powerful patents: No. 5,843,780 and No. 6,200,806 that cover all human embryonic stem cells and the method by which they are made. Investigators have to pay WARF patent fee of $5,000 to obtain permission for the derivation and use of stem cell lines. In case a scientist develops a research tool, a therapy or some other useful invention originating from embryonic stem cells, WARF can interfere and claim to share the commercial rights. The monopoly WARF is holding over all future research because the patent is broad and covers all human embryonic stem cell lines in the U.S., not just the specific lines developed by Thomson (Washburn 2006).

Thomson claimed that he created five stem cell lines by destroying thirty six frozen or fresh embryos (Thomson et al. 1998). Dr. Lanzendorf (2001 Michigan) and his team have had 101 embryos available to derive three lines. He stated that his work was done under optimal circumstances. It is more difficult than it sounds, because frozen embryos rather than fresh embryos reduce the success rate to obtain a blastocyst. On average thirty embryos are needed to establish one cell line (Graf 2003). Species specific differences may explain the long delay in transferring the results from mouse to man. The proper media and reagent for optimal culture of the harvested stem cells were unknown. In 1990 for the first time, non-human primate stem cells were isolated at the National Primate Research Center of the University Wisconsin. Through these experiments researchers tried to define the best-suited culture media (Thomson 1996, Kiessling 2010). However, scientists have not managed this challenge yet; because of the ability for cells to proliferate in these currently developed media is less than one percent. Steady culture conditions were widely used for a long time. During the propagation in the cell culture arise chromosomal, genetic and epigenetic modifications. It is assumed that embryonic stem cells are being manipulated by in vitro culture conditions. In this regard, it is impossible to use stem cells for transplantations, even assuming that the pathway of differentiation is available. Until today, not even one study was undertaken between the linkages of an imprinted expression of culture media unto stem cells (Doherty 2000). Scientists are searching for the prime quality of culture media. It is desired that human embryonic stem cells maintain all embryonic stem cell features in the culture, which are pluripotency, immortality, and the ability of continuously culturing them in an undifferentiated state. Researchers are faced with a major workload to modify the proper media conditions for human embryonic stem cells. Research is conducted to examine whether mutation affects cell-proliferation (Verlinsky 2005).

9.4 Adult stem cells

In Germany, research priority is given to adult stem cells. German scientists claim to be suspicious of human embryonic stem cell research because the lack of results on cell based therapies. In Germany twenty one embryonic stem cell lines were available. In April 2008, the German parliament allowed the use of stem cell lines created in foreign countries before May 1, 2007, thus, increasing the lines up to 500. Nevertheless, federal funds are going mainly into adult stem cell research (DFG 2006). Adult stem cells are undifferentiated cells found in each tissue of the human body. Multiplied by cell division, adult stem cells replenish dying cells and regenerate damaged tissue. Scientific interest in these cells has centered on their ability to generate cells from the organ from which they are originated. Many studies and even successful therapies in humans were accomplished. Bone marrow cells have proved their plasticity and are in clinical trail studies. Adult bone marrow stromal cells are already converting into blood cells. Medical stimulation is responsible for an increased production of bone marrow cells in sick or healthy people. Adult stem cells can be isolated easily from donors or patients. After genetic manipulation, they can be given back to the patient. Thus, a therapy concept is well known from bone marrow donation or transplantations. Chemotherapy destroys normal blood-producing stem cells in the bone marrow. The blood stem cells from the patient, or the donor, collected before the chemotherapy will improve the treatment. After the high-dose chemotherapy, the stem cells are infused into the patient's bloodstream to restore blood cell production. Stem cells can be stored by freezing until they are needed again. Cardiologists state that we are only beginning to discover what adult stem cells can accomplish. Adult stem cell therapies currently treat congestive heart failure, cardiomyopathy, peripheral artery diseases and many more.

In 2001-2002 researchers from Minnesota and Yale were able to attain adult stem cells proliferating extensively without obvious senescence or loss of differentiation potential. They may be an ideal cell source for therapy of inherited or degenerative diseases. One can observe up to eighty cell dividing cycles in these cells. The plasticity of the cells differentiated from adult progenitor cells to the hematopoietic lineage and additionally to the epithelium of liver, lung and gut (Jiang et al. 2002). Many exciting discoveries helped to understand the adult stem cell plasticity. They are not only capable of producing functional cells of one organ system, they also have the flexibility to differentiate into multiple other cell types. Bone marrow cells produce hematopoietic stem cells. They also have the potential to differentiate into mature cells of heart, liver, kidney, lungs, skin, bone, muscle, cartilage, fat, endothelium and brain (Kraus 2001). Clinical trials have hinted at adult stem cell therapy's tremendous potential. A study assessing the acute and long-term effects of intracoronary adult stem cell transplantation demonstrated that injecting 6.6×10^7 bone marrow cells into damaged hearts improved ventricular performance, quality of life and survival. Still scientists do not understand how these cells work their therapeutic magic (Strauer et al. 2010).

Biomaterials replacing the bone structure are more and more desired. The age of the population is increasing. The European Union has an annual increase of 750,000 hipbone replacement operations. Five hundred thousand knee operations are indicating that bone replacement materials are needed to adapt to the natural occurring remodeling of bones. Bone tissue-engineering benefits from stem cells that cover calcium phosphate surfaces. The bio-adhesive surfaces promote osteoblast differentiation (Müller et al. 2008). The US Patent 6200606 from March, 2001 was given to isolate precursor cells from hematopoietic and non-hematopoietic tissues. They were used in vivo for bone and cartilage regeneration. The understanding of the physiologic processes in the bone growth plate and the regenerative potential of cell proliferating chondrocytes has already modified the clinical therapy of growth plate fractures in juveniles (Breburda et al. 1996, 2000 and 2001 a and b). Other sources to repair bone or cartilage damage are: peripheral blood and adipose tissue and a population of cells isolated from marrow, which do not require in vitro culturing (Williams et al. 2001).

In 2001, it was assumed that only human embryonic stem cells have plasticity faculties. Are adult stem cells competing with human embryonic stem cells and are we too euphoric about embryonic stem cells and simply disregard the capacity of adult cells? The fact is that adult stem cells are successful therapeutics. "Surgeons now start to see and understand the very real potential for adult stem cells and tissue engineering to radically improve their ability to treat patients" says Baroness Morris of Bolton in a debate of Britain's second parliamentary chamber in 2009.

10. "Ethicizing" of technology conflicts, aspects contra dilemma

10.1 Interspecies conflict

Some call it a mismatch made in heaven. It all started out with a very lonely Philippine stallion who was isolated by other horses. "He fell in love with a herd of zebras," reported the Manila City Zoo official on August 16, 2010 who gave an interview about "his" week-old offspring "hebra", a crossbreed of zebra and horse. The mismatch of genes from those two animals will most likely have the foal that develops several complications. "A very rare occurrence has happened and this is probably the first hebra in the country," assured Dr. Donald Manalastas the division chief of the Zoo. Zebras, horses and donkeys can interbreed because they come from the same equus genus. A zorose is a cross between a zebra stallion and a horse mare and a zonkey that has a zebra father and a donkey jenny. The Manila Zoo has had two hebras in the past. But they died just after they were born, due to certain genetic complications (Postrado 2010). Hybrids of two species are observed in nature. On May15, 1985, a similar incident happened when trainers at Hawaii Sea Life Park were stunned about their gray bottlenose dolphin named Punahele. She gave birth to a dark skinned calf that resembled the 2,000-pound male false killer whale. Both shared the same pool at the Park.

In 2006 a Canadian Arctic Hunter shot a bear that resembled a polar bear with a grizzly bear's features. Scientists estimate that hybridization among distinct species is not so rare. It is not some kind of moral breakdown in the animal kingdom. Nevertheless, it is thought that ten percent of two animals of the same species interbreed and that it occurs occasionally in twenty-five percent of plant species liaisons. The vitality of the hybrid will determine the long-term success of such produced offspring. Several human-bred animals by artificial insemination are well known such as beefalo, a bison-beef cattle, which is usually infertile (Carroll 2010). Sterile animals frequently arise through hybridization, because of a link among the genetic phenomena of asexuality. When the two species are genetically very distant, or carry different numbers of chromosomes, the offspring are infertile and usually unviable. From this prospective genetic engineering is a dead-end road. "We are breeding animals for big chests, and our turkeys are not able to reproduce by themselves anymore," is a common explanation from food scientists (Ritter 2010).

In 2004 the mystery of the monster-hog arose from the swamps and woods of Georgia. Chris Griffen, a hunting guide, discovered and killed the animal resembling a hog. People were looking for a scientific investigation to find the real roots of the record-size boar. The twelve-foot, 1,000-pound wild hog, called Hogzilla, was exhumed for a DNA test to reveal, if possible, that the record-size boar is a "pig in a poke". Is Hogzilla real? National Geographic says so. It was determined that Hogzilla was a hybrid of wild boar and domestic swine. From whence he came is not clear (Dewan 2005).

The most beneficial hybrid in modern society is the genetically engineered silkworm. They spin like spiders and create a fabric that could be used in everything from bulletproof clothing to artificial tendons. The hybrid of silkworm silk and spider silk, called spider-silk-spinning silkworms could shake up the textile industry. Silkworms "clothed" peoples for thousands of years by producing large quantities of material. Spider silk is significantly stronger but hard to make. Spider silk proteins have already been produced in plants, in bacteria and even in goat's milk. Still, the protein differs from spun spider silk. Experts consider that silkworms have the necessary body parts to spin the protein into silk threads. Experts at Notre Dame and University of Wyoming are now planning to replace multiple silkworm silk-producing genes with spider silk genes. The only deficit of hybrids will remain that they cannot reproduce (Bland 2010).

Reproduction is fundamental to all known life. The biological process produces individual descendants that have a combination of genetic material contributed from usually two different members of the species. Each offspring exists as a result of reproduction wherein they might have come into existence by sexual or asexual means. The primary form of reproduction is asexual. Single-celled organisms, such as bacteria reproduce without fertilization. To reproduce without the opposite sex of that species is not limited to primitive organisms. Parthenogenesis is a way of asexual reproduction. Parthenos means "virgin", genesis comes from "creation". It describes the growth and development of an embryo or a seed without fertilization by male. It occurs naturally in invertebrates, and vertebrates, in some reptiles, scorpions, fish and very rarely in birds and sharks (Eilperin 2007). Parthenogenesis is very rare and most animals capable of asexual reproduction are very simple. The weakness of parthenogenesis is seen in a low genetic diversity susceptible to harmful mutations that persist through generations (Anissimov 2010).

Strongyloides stercoralis, also known as threadworm or human parasitic roundworm causes the disease of strongyloidiasis in humans, dogs and cats and other mammals. Parthenogenetic females produce 2000 eggs daily in the host. The nematode has a heterogenic life cycle, which consists of a parasitic generation and a free-living generation. The free-living males and females of the threadworm die after mating. The infectious larvae penetrate the skin of the host. Some of them enter the superficial veins and travel through blood vessels to the lungs to enter the alveoli. Coughed up and swallowed they finally mature in the small intestine. Only females will reach reproductive adulthood in the intestine, where they reproduce parthenogenetically. Young larvae depart the host in the natural way. Pneumonia-like symptoms might lead to a misdiagnosis and treatment with cortisone. The cortisone-steroids are similar to a growth hormone of the parasites and increase the reproductive capacity of the parasite inside the host (Siddiqui et al. 2000). Ironically, the treatment results in adverse effects due to unawareness of the true cause.

There are no known natural cases of mammal parthenogenesis and it has never been observed so far, either in humans or in mammals. Gregory Pincus induced parthenogenesis artificially in rabbits in 1936. In 2002, primates were generated by parthenogenesis (Cibelli et al. 2002). The Tokyo University of Agriculture produced monkey and mouse offspring from noninseminated egg cells in 2004. Abnormal development results when the mother's genomes are imprinted twice in the offspring. Many mammals' genomes are completely dependent on a mix of genes from both sexes. Due to ethical reasons it is unlikely to pursue human live births from parthenogenesis, whereas scientists assume that the process can be used to create embryos for experimentation (Anissimov 2010).

Parthenogenesis is somewhat familiar to us in connection with the South Korean research scandal of Dr. Hwang Woo-Suk. We might remember him, especially through his deceptive apology. Hwang originally claimed that he and his team had extracted stem cells from cloned human embryos. However, Hwang unknowingly produced the first human embryo from parthenogenesis (KBS 2006). In Hwang's case inaccurate techniques led to an embryo created without fertilization. His lab enucleated hundreds of egg cells. Normally the primordial diploid egg cell becomes haploid shortly before fertilization. From the two polar bodies, one set of chromosomes will disappear. But if the ejection is accidentally prevented an outcome similar to parthenogenesis will occur. Hwang attempted to be the first to clone a human. But instead of successful cloning, he produced viable human embryos obtained by artificial parthenogenesis.

A parthenogen always yields females, because no sperm is involved. Parthenogenesis is distinct from artificial cloning where the organism is genetically identical to the donor. The parthenogen is like a genetically unique sibling to the mother.

Until now, it is not clear whether stem cells, derived from therapeutic cloning or parthenogenesis, will be suitable for research because mammals have imprinted genetic regions. Imprinted means that some of our genes come from one parent to work normally. These genes have the ability to be turned off or on, depending on which parent contributed the gene (Keder 2004). Either the maternal or the paternal chromosome precedes a normal development in the offspring. A double dose of maternally imprinted genes leads to abnormalities. Gene signaling is one of the focuses of research exploring the secrets of embryonic development, which means also that the mechanisms of gene signaling and regulation are helpful to gain insights for stem cell differentiation. In 2007 scientists from the Jacobs University in Bremen, Germany and colleagues from Emory University, Atlanta, Georgia discovered a pattern in the DNA sequence of the mouse genome that may play a fundamental role in the way DNA molecules regulate gene expression. Most gene expression is based on the two copies every living being has inherited from the mother and from the father. Only a small subset of genes is allele-expression specific, which means either the copy of the father or mother is "turned on". The other gene is silenced or turned off and will not manufacture proteins. Researchers found out, that so-called epigenes are just as important as the genetic sequence itself in order to determine gene functioning. They know that DNA methylation, a biochemical reaction that adds a methyl group to DNA is one of the epigenetic processes that turns genes off. How and when the process is accomplished for either silencing or expression is still a mystery (Da Jia et al. 2007).

Over the past few years, new approaches have been proposed to avoid the discussion with those who object to the destruction of human embryos. To create abnormal stem cells from parthenogenetic embryos is not acceptable either ethically nor scientifically. Nevertheless, in 2008, researchers from the Department of Reproductive Medicine, at the University of Buenos Aires, Argentina, reported that they were able to develop parthenogenetic blastocysts after artificial activation of noninseminated human oocyte after thawing. That means that in this instance scientists have knowingly induced parthenogenetic embryos. To the astonishment of the researchers thirty-six of thirty eight cryopreserved noninseminated human oocytes survived the thawing process. Five of the artificial- activation-produced parthenogenetic embryos reached the blastocyst stage (de Fried et al. 2008). Scientist speculated that in the future the procedure of parthenogenetic embryos may provide a source of human embryonic stem cells. Still the method remains of questionable scientific validity, because paternal genes are missing (Zinkant 2007).

10.2 New discoveries that really are not

The mention of stem cell therapy draws reactions. It is a topic most people have heard about by now. But how much knowledge do people have? Still, the concept is likely to be foreign to most. What might come first to our minds are human embryonic stem cells that are harvested from human embryos, which are destroyed in the process. Or we might have heard about fetal stem cells coming from abortions or miscarriages. For the most practical reason, fetal stem cells are not able to be used as a cure. Fetal stem cells divide so rapidly that they are excellent producers of their own characteristics, namely cancer. The risk is very high which makes them largely unsuitable for therapeutic uses. In Germany, Mexico and Panama, stem cell research is based on the use of adult stem cells, harvested from the patient's own bone marrow and blood. They are far from all the ethical constrictions and have resounding success. The list of diseases being routinely cured with adult stem cells is impressive and extensive. Diseases formerly thought to be incurable are Amyotrophic lateral sclerosis (ALS), Alzheimer, Autism, Cardiovascular Diseases, Cerebral Palsy, Diabetes, Failed Back Surgery Syndrome, Macular Degeneration, Multiple Sclerosis, Osteoarthritis, Rheumatoid Arthritis, Parkinson Disease, Spinal Cord Injuries, Stroke and many more. Solid science justifies the use of adult stem cells.

All life starts with fertilization of an ovum by sperm. When the blastocyst implants into the womb, the continuation of the development is granted. Trophoblast cells from the blastocyst invade the uterus wall. They rapidly grow in order to prepare a place for the embryo to attach itself. The embryo will develop from the stem cells of the embryoblast and eventually the differentiated cells take over a specific function of the body. Even so, throughout our lives we have cells that remain very unspecialized. They can continue to divide and help our body's organs to regenerate and repair themselves. These cells are called adult stem cells. They have been found in every body tissue. Modern adult stem cell therapy tries to exploit this process (Knoepp 2009).

Whereas, human embryonic stem cell research focuses only on the embryoblast, the inner cell mass, where the unspecialized human embryonic stem cells are located. By destroying the blastocyst to harvest the cells, the source of nourishment for the embryoblast is destroyed. Prior to implantation the trophoblast (ring of flattened cells) nourish human embryonic stem cells. Later on, the placenta provides oxygen and nutrients for the proper embryonic development. Detached from trophoblast, human embryonic stem cells do not even start to specialize into all the cell types of the adult organism.

Science is far from being able to differentiate human embryonic stem cells in a simple Petri dish. Besides, experiments performed in test tubes differ substantially from the natural development occurring in the body. In the beginning, the broad developmental capacity of embryonic stem cells was studied in the culture dish, where undifferentiated spontaneous differentiation occurred, called embryonic bodies. Then embryonic stem cells were injected in a living vector (mice) and teratomas established. Both objects have a tumor–like formation, containing tissue seen in very early embryogenesis. But direct differentiation of human embryonic stem cells into functional tissue has never been observed. Scientists employed a whole-body (mice) model to observe the conservation and recognition of common signals for neural differentiation. In a chimeric model, human embryonic stem cells were implanted in the brain ventricles of embryonic mice. The human embryonic stem cells "adapted" to the live environment and differentiated into mature human neurons. They fully integrated into adult mouse forebrain. It is anticipated those kind of studies will generate a new way to study human neurodegenerative and psychiatric diseases. Or at least it is thought that such a model could speed up the screening process for therapeutic drugs (Muotri et al 2005). The cell mixing experiments with human embryonic stem cells in mice demonstrated that they engraft into mouse brains where they proliferate, differentiate and persist. It is also common to inoculate human embryonic stem cells into mouse blastocysts and implant the embryonic chimeras in the uterus of foster mice. The mouse embryo can be used as a surrogate for human embryonic stem cell differentiation (James et al. 2006). It looks like scientists simply let mice do the job of specializing human cell types in a non-human animal model.

The Ramsey Colloquium, a group of Jewish and Christian theologians, philosophers and scholars, argue that:

"embryonic human life is treated simply as research material to be used and discarded and should even be brought into being solely for that purpose. It is morally repugnant, entails grave injustice to innocent human beings and constitutes an assault upon the foundational ideas of human dignity and rights essential to a free and decent society. The request to destroy human embryos for research purposes because they are unfit or are leftovers from in vitro fertilization is unconscionable and would cross the threshold into a world of limitless technological manipulation and manufacture of human life" (Ramsey Colloquium, 2002).

In countries like Germany, cloning and the production of embryos for research purposes is forbidden, even in 2010. Whereas, elsewhere, scientists at the Jones Institute for Reproductive Medicine in Virginia announced on July 11, 2001, that they intentionally created human embryos from donor eggs and sperm. The sole purpose was to conduct "destructive research". They argued they do not think it is justified to do research on so-called "spare embryos" left over from in vitro fertilization procedures. One hundred and sixty-two egg cells obtained from donors were fertilized. Fifty embryos reached the blastocyst stage in order to generate three stem cell lines. The researcher got the permission and support of the ethics committee of the former mentioned institute, despite the fact of strict regulation regarding the creation of embryos for research (Graf 2003). According to the journal Fertility and Sterility, the official publication of the American Society for Reproductive Medicine, the ethics committee of the Virginia Institute announced: "The creation of embryos for research purposes was justifiable and that it was our duty to provide humankind with the best understanding of early human development." Ethicists started to ask why the understanding of human development requires the destruction of the developing humans? And what ethical theory permits the destruction of one smaller group of human beings for the benefit of a larger, more powerful group of human beings? "It looks like the greatest good for the greatest number of people, the lives of a minority may be sacrificed for the benefits of the majority. But what can be a more vulnerable minority than human embryos, the tiniest members of the human community who deserve at least a right not to be unnecessarily harmed? They are genetically unique individuals of the species Homo Sapiens, just like you and me," argues Dr. Mitchell from the center for Bioethics and Human Dignity, at the Trinity International University in Illinois, USA. Mitchell is convinced that: "we cannot open the moral floodgate to permit human embryonic stem cell experimentation, because we will let in a host of moral evils." He predicted that there will be no turning back. "Now embryos are being created for the purpose of research. Next, we will be told that quality control demands that we clone those embryos to ensure that they are all identical. Then, we will be informed that, in order to learn more about human development, of course, we need to bring a human clone to term" (Mitchell 2001).

In February 2004, Dr. Woo-Suk Hwang, a Korean veterinary doctor, made world headlines when he claimed to have cloned human embryos. He used the technique of somatic cell nuclear transfer. Later on he asserted the establishment of eleven patient specific stem cell lines by using 185 eggs. Laboratory notes indicated that he used at least 273 eggs. To determine exactly how many eggs were involved is impossible. From November 2002 to November 2006, Dr. Hwang collected a total of 2061 eggs. Dr. Hwang stated that he was unaware of egg donations by laboratory members, whereas graduate students who donated eggs got approval from Dr. Hwang.

For therapeutic cloning three steps are necessary: nuclear transfer to create an embryo, the blastocyst formation and the generation of the stem cell lines. The patient specific disease carrying DNA has to be repaired before the nuclei can be transferred. Afterwards it is hoped the stem cells will differentiate into the desired cell types. They also must be free of tumorigenic potential. Dr. Hwang's technique had been used long before in veterinary medicine. His article in Science let cloning proponents declare that the age of therapeutic cloning is reached and soon they predicted sick patients would be able to get their own "tailored" embryonic stem cells to cure them. Also, genetically matched stem cells could be derived for use in regenerative medical treatments. Hwang's paper was very well received. His "breakthrough", accomplished in South Korea, proved that the United States did "fall behind" in stem cell research. More than ever people blamed President Bush's policy for the failure. As a consequence, his limiting federal funding of embryonic stem cell research should be overturned to permit American research to flourish. Hwang was lauded internationally as a genius and national hero. Until Gerald Schatten of the University of Pittsburg complained that Hwang had purchased the human eggs he used in the experiments. By that he violated the ethical canons required that they be donated. Shortly after more rumors followed. It was obvious, that Hwang had committed research fraud. Hwang forced a junior researcher to submit wrong photos to the Science magazine. Science coauthors stated that nine of eleven stem cell lines were faked. The DNA of the eleven lines did not match the patient's DNA. Prof. Hwang's research team did not create patient-specific stem cell lines. He did not use therapeutic cloning; instead, he fertilized the oozytes. South Korean scientists discovered Hwang's missteps, whereas the more experienced peer reviewers for Science failed to do so. Nevertheless, the media pursued the story with praise until it was revealed as bogus (Smith 2005 and Seoul National University's report on Dr. Hwang Woo Suk 2006).

Until November 2005, Dr. Hwang was considered as the pioneering expert in the field of stem cell research, best known for his two publications in Science. His fraudulent claim of having successfully cloned human embryos cost him his reputation, job, financial and legal support. He was charged with embezzlement and bioethics law violation. He failed to provide scientific proof and verifiable data for the research. Despite the notable absence of any data, the media and public celebrated Dr. Hwang and his publication as a breakthrough. Hwang made 2005 headlines when he criticized U.S. President George W. Bush's policy on human embryonic stem cell research. Dr. Hwang raised 500 Million US Dollars as donations for his research. He promised to clone a BSE resistant cow, a Siberian Tiger and animals that are on the list of endangered species (Koreans Report 2006).

The South-Korean President Roh Moo-Hyun defended Hwang. Roh stated the bioethical discussion is up to the politicians and should not be transferred to researchers. Roh reaffirmed Hwang with unflinching support. Roh hoped that the researcher would return to his lab soon. Furthermore, Hwang should not be pulled into ethical discussions (Korea Net news, 2005). Hwang stated that he was not aware of the Declaration of Helsinki. The World Medical Association as well as the Nürnberger Code and the Hippocratic Oath to ensure medical ethics principles developed the declaration. Hwang apologized for creating a shock and disappointment. He stated: "I was blinded by work and my drive for achievement." He put the blame on his contributors for having deceived him with false data. Hwang insisted that if he would get six more months he could prove that he has the technique to clone humans (ABC News 2006).

Does Dr. Hwang consider himself a prisoner of his own conscience? Can researchers make use of the conscience clause, or would they lose their jobs, or university or college positions by doing so? During the Nazi regime in Germany, Jews were prevented from achieving higher academic degrees. Later on, the communist ideology in the East German Socialist Republic decided who was allowed to study in which place and on which subject. Even so, if the parents did not comply with the pattern, highly intelligent children were rejected. In the present day, students who are against the use of animal models are at risk of not making their grades or getting credits. The use of the conscience clause is not accepted at all and is viewed as counterproductive. However, who is protecting animals, especially monkeys that are used as research subjects? It is relatively new that newspapers are writing about the protecting of our fellow creatures and the conflict with experimental animal research (Wandtner 2008). Already Engedahl observed that researchers must fear the loss of their jobs if they protest against agricultural companies. What was left is that Hwang became a pariah; he had the intention of proving that therapeutic cloning is not science fiction. Other scientists tried as well to grow a new population of patient's own cells. Building up such a stockpile, however, has proven practically impossible. Nobody considered the very questionable benefit of inserting a defective somatic cell into an enucleated egg cell. The necessary genetic manipulation of the DNA prior to insertion remains a difficult undertaking and might be an elusive goal.

Summa summarum, we don't know whether human cloning has been successfully accomplished or not. We also lack proof whether embryonic stem cells have been derived from cloned embryos. We don't even have the knowledge to what depths the dishonesty of the seemingly most successful researcher in the field actually descended, concludes Wesley Smith, the author of the book <u>Consumer's Guide to a Brave New World</u>. Smith further stated: "For years, human cloning has been promoted through propaganda techniques of misrepresentation, exaggeration, and false hope for the suffering. Instead there is the ongoing hype about the medical potential of cloning. It reached cruel heights in the wake of President Reagan's death from Alzheimer disease." On June 10, 2004, Rick Weiss, a Washington Post science reporter revealed in his article: "Alzheimer disease is extremely unlikely to be effectively treated with human embryonic stem cells, no matter whether they are coming from clones or are harvested from destroyed young human beings." Weiss continues that a stem cell expert confessed: "of all the diseases that may someday be cured by embryonic stem cell treatments, Alzheimer's is among the least likely to benefit. Alzheimer is a whole brain disease that involves the loss of huge numbers and varieties of the brain's 100 billion nerve cells and countless connections, or synapses, among them."

But why do proponents of human embryonic stem cell research go on to invoke a cure of Alzheimer's? Weiss thinks Ronald D. G. Mackay, a stem cell researcher at the National Institute of Neurological Disorders and Stroke might have given a proper answer in an article at the Washington Post some years ago. Mackay said: "To start with, people need a fairy tale. Maybe that's unfair, but they need a story line that's relatively simple to understand. Human embryonic stem cells have the capacity to morph virtually any kind of tissue, leading many scientists to believe they could serve as a universal patch for injured organs" (Smith 2005).

On October 11, 2010, the day the US celebrates Columbus Day, the Shepherd Center, a 132-bed spinal cord and brain injury rehabilitation hospital and clinical research center in Atlanta, Georgia participated in an experiment for restoring sensation to people with spinal cord injuries. They cooperated with Geron Corporation, a California based firm that conducts "dubious embryonic stem cell research". The firm announced it has treated the first time patient with embryonic stem cells. A human being had been injected with human embryonic stem cells, coming from human embryos left over from fertility treatments, explains Dr. Okarma, CEO Director of Geron. Seven potential sites in the United State enrolled patients in the clinical trial. Geron Corporation got the first U.S. Food and Drug Administration license to use the controversial cells to "treat" people. The patients eligible for the phase I trial are those who have suffered a complete thoracic spinal cord injury no longer than seven to fourteen days ago. Patients in such conditions can still move their arms and breath on their own.

However, the injected embryonic stem cells aren't pure human embryonic stem cells anymore. The patient received precursor cells, un- or only partially differentiated cells, derived from human embryonic stem cells. It is possible to make an abundant supply from the undifferentiated embryonic stem cells before treatment of an acute spinal cord injury. The hope of the Geron Cooperation is, that the cells once inside the human body will finish specializing, then travel to the site of the recent spinal cord injury and release compounds that will repair the damage. The whole trial was never set up to cure patients. It is only used to determine that the cells are safe to use. It is actually no more than a safety test to see whether the injections of precursor cells lead to tumors or other complications. And it will take some time, perhaps many years. Still no one can be sure whether the stem cells really helped. Meanwhile researchers who are looking at the facts are convinced that adult stem cell therapies make experiments with embryonic stem cells wholly superfluous (Smith 2010).

Scientists expressed concerns about Geron's trial. The cells used were not completely differentiated, only partially differentiated ("precursors"). They actually did not inject human embryonic stem cells because science is still not able to overcome significant problems in their use with animals. Specifically, undifferentiated or partially differentiated embryonic stem cells cause tumors and are rejected by the immune system.

Researchers use different culture conditions in order to induce neuronal differentiation. They add Vitamin A, or use a so-called co-culture method. That means human embryonic stem cells are cultured in the Petri dish with already differentiated neuronal tissue. "However, the neural induction process often requires direct contact of human embryonic stem cells with the stromal cells (connective tissue cells of an organ in order to release growth factors that promote cell division), which introduces unknown and potentially risky factors as most of these cell lines are tumorigenic," explains Su-Chu Zang in his article: "Human embryonic stem cells for brain repair," published at Biological Science 2007. It is also common to isolate neural stem cells or progenitor stem cells from fetal brain tissue obtained from fetal tissue banks, which are using the remains (placenta, extra-fetal tissue....) of aborted children for medical research.

Former Indiana State University Biology Professor Dr. David Prentice accused Geron Corporation of falsely promoting its work as embryonic stem cell trials when derivates were used instead. "Geron is irresponsibly trying to do science by press release, publicizing that they have begun their human experiments by injecting a patient with potentially dangerous cells. Their press hype will help stock prices, but not science and not patients. We hope the patients don't suffer any ill effects. It will be years before there is hard evidence about safety or effectiveness. If Geron truly wanted to help patients, it would stick to research and trials involving adult stem cells. They already are helping patients now who are battling dozens of diseases and medical conditions." Prentice continued: "Adult stem cells have published real scientific evidence for effectively treating spinal cord injury."

Professor Dr. John A. Kessler, chairman of neurology and director of the Stem Cell Institute at Northwestern University announced that the first application from Geron for the embryonic stem cell trial was flawed. Already in 2008, Evan Synder, a neuroscientist, head of the Stem Cell Research Center at Burnham Institute for Medical Research in San Diego, warned that the research is not ready for humans. About the Geron trial Synder noticed that the mice used to conduct pre-human trial research had excessive injuries. And the company should have done experiments involving larger animals before seeking FDA permission. Did Geron move too fast? Jerry Silver, a neuroscience professor and stem cell researcher at Case Western Reserve University in Cleveland is absolutely convinced they did. He is not alone with that view. "Nobody can conceive of a human trial with the use of human embryonic stem cells following immediately from experiments in rodents only," explains Jerry Silver. Many spinal cord researchers acknowledge that the pre-clinical data itself doesn't justify the clinical trial. They know too well that science revealed that many treatments actually worked well with rodents, (or even only in rodents) but failed miserably in humans (Ertelt 2010).

10.3 Future of regenerative medicine

Researchers discovered in bone marrow two kinds of adult stem cells in the 1950's. Ten years later scientists found in the brain two regions that contained dividing cells, which became nerve cells. Until the 1990's the scientific community believed that the adult brain could not generate new cells. In adult organisms, stem cells and progenitor cells work together as a repair system for the body. They maintain the normal turnover of regenerative tissues such as blood, skin, or intestine, but also replenish specialized cells. Adult stem cells reside in a specific area called a stem cell niche. They enter normal differentiation pathways to form the specialized cell types of the tissue in which they reside. When needed they are able to divide and can give rise to mature cell types that have the characteristic shape, specialized structures and function of the particular tissue.

Scientists are just beginning to understand the mechanisms underlying cell differentiation. The future of regenerative medicine is based on replacing damaged cells and repairing deficient organs by the use of stem cells. Stem cells definitely have the capacity to proliferate indefinitely. Theoretically, the supply of all types of cells to the human body is possible. The processes to reprogram adult cells to embryonic cells are complex and full of risk. The same can be said when researchers try to retransform embryonic stem cells into adult cells. Obviously questions arise such as whether it might be simpler to generate the needed cells without passing through the embryonic pluripotent stem cell stage?

In 2009, Science reported about a study from a research team at the Centre d'Immunologie de Marseille Luminy, France led by Professor Dr. Michael Sieweke. They achieved the regeneration of macrophages without the reversal to the embryonic stem cell intermediates. Macrophages are specialized cells in the immune system, which normally digest ("eat") bacteria or cell debris. As soon as a cell has acquired a specialized function, like a mature brain neuron cell, a mature muscle cell, or a macrophage for the immune system, etc. they cease to divide and remain "blocked" in this state until they die. The Marseille team genetically modified mouse macrophages that are able to reinitiate cell division of specialized cells. Reinjected into the animal, these modified cells behaved normally, did not form tumors and perfectly performed the tasks expected of them. The adult macrophage ingested bacteria and secreted the chemical agents capable of killing them. These findings suggest that specialized cells can directly be reprogrammed to immature adult cell types. The passage via embryonic stem cells may not be necessary to enable the regeneration of cells to repair damaged tissue (Aziz et al. 2009).

Adult stem cell research is rapidly proceeding to uncover the general molecular mechanisms that control self-renewal and differentiation. More and more experiments show that certain adult stem cell types can differentiate into cell types seen in organs or tissues other than their own. Scientists speak about a transdifferentiation. For example brain stem cells differentiating into blood cells. The above-mentioned experiment of Professor Sieweke used a well-controlled process of genetic modification to "reinitiate" cells. A strategy that will offer a way to reprogram available cells and to replace them with cell types that have been lost or damaged by a disease. In type I diabetes, or juvenile diabetes, pancreatic insulin producing beta cells are lost. A mouse experiment showed that researchers were able to create beta cell-like cells able to secrete insulin. The reprogrammed cells were similar to beta cells in appearance, shape and expressed genes characteristic of beta cells and could partially restore blood sugar regulation in the animal model (Zhou et al. 2008).

Two years later, Pedro Herrera, professor in the Department of Cell Physiology and Metabolism at the University of Geneva Medical School, exposed rodents to a toxin that destroyed more than ninety-nine percent of beta cells only. The researchers labeled the alpha cells so they could track those cells. Alpha cells secret a hormone that counteracts the effects of insulin producing beta cells. Mice were given insulin therapy to keep them alive. Alpha cells spontaneously changed into functioning beta cells and insulin therapy was no longer needed. Their research published on April 22, 2010 in Nature is the first to show that beta-cell regeneration was derived from alpha-cells. It revealed that the pancreas cells still have a certain degree of cell plasticity and change into insulin-producing beta cells (Thorel et al. 2010). Diabetes experts cautioned. "A number of mouse findings haven't translated well in human research," argued Dr. Davie Kendall, chief scientific and medical officer at the American Diabetes Association. In type I diabetes, the immune system attacks beta cells. This process will not stop. People who received transplants of insulin-producing beta cells needed to go back on insulin, simply because the immune system destroyed the transplanted cells. "Any time you are thinking about any type of cure or really good treatments for type I diabetes, you have to consider both, the beta cells and the immune side" concluded Andrew Rakeman, the scientific program manager in beta cell regeneration at the Juvenile Diabetes Research Foundation (Gordon 2010).

Researchers want to learn more about the indefinite self-renewal of adult stem cells. Cardiologists increasingly rely on adult stem cells. In 2007 the Houston University of Texas, Department of Cardiology reported an animal experiment in which cell-fusion occurred between adult blood stem cells (hematopoietic progenitor cells) and heart muscle cells (cardiomyocytes). The fusion is possible when cell adhesions proteins that bind the two cell types together are available. Once they are fused they divide again to produce enough cells to help the heart to contract. The study's lead author Professor Yeh explained: "The accepted dogma is that heart cells cannot divide, but we show that fusing adult stem cells into muscle cells bestows these cells with a new and wonderful ability to divide again to produce new muscle cells (myocyte) to repair the heart. It is marvelous that adult stem cells can help heal the heart. By understanding the mechanism involved, we may be able to refine and optimize the process," Dr. Yeh says. Unfortunately there are not many stem cells available. Scientists from St. Luke's Episcopal Hospital in Houston and Dr. Yeh's University team are focusing now on boosting the response, to treat a heart attack more effectively. According to what they have already learned about how adult stem cells can build, it may be possible to choose between rebuilding the heart muscle or restoration of blood vessels. It all depends on what is therapeutically best for the patient (Zhang et al. 2007).

10.4. Umbilical cord stem cells

The use of adult stem cell research has been performed for sixty years (until 2010). In the US the earliest adult stem cell transplant used bone-marrow cells. The treatments of blood-borne diseases are addressed by hematopoietic (blood) stem cells from bone marrow, peripheral blood and human umbilical cord blood. Bone marrow donors must undergo a surgical procedure. Umbilical cord blood stem cells are collected from the umbilical cord at birth. These cells can produce all of the hematopoietic (blood) cells. The major use of stem cells for reparative medicine was seen in treating cancer patients, because chemotherapy destroys the bone marrow. At present, we know that umbilical cord blood stem cells can differentiate into various non-blood cell types, including neural cells. Several animal models indicated the successful use for treatment of neurological injuries. The United States is, for some people, too restrictive in using the benefits of human umbilical cord blood based cures for diseases. Consequently some people are heading overseas. There are clinics around the world that offer treatments for medical conditions. Who wonders that society is not troubled in looking for shortcuts no matter whether it is for medical or reproductive reasons. Still the most attracting medical tourism is for human embryonic stem cell therapies. There are widely publicized success stories of "miracle" cures with the controversial therapy. Scientists are afraid that people may harm themselves. The use of embryonic stem cells carries serious and unpredictable medical risks. Researchers warn that no reliable evidence is given that any of the costly treatments work (Lord Alton 2001). Still thousands of people are putting their health and life savings at risk for these unlicensed treatments. It seems that nothing will stop them from doing so. Ethicists like Wesley J. Smith wonder why superior therapies are pushed away, while dubious announcements about human embryonic stem cell treatments are getting media attention instead. Wesley recently said: "The mainstream press is so intoxicated with embryonic research, they virtually ignore non-embryonic breakthroughs" (Smith 2005)

Many believe that insufficient attention has been paid to the potential of adult stem cells as compared to the exploration of umbilical cord stem cell research. It is known that umbilical stem cells have no surface antigens to cause reactions. Allergic responses that occur with embryonic stem cells are not common. Cord blood and other forms of adult stem cells use is allowed to treat a number of diseases and conditions in the United States.

However in 2005, the FDA has not approved or given consent for physicians to utilize cord blood or cord blood stem cells to treat neurological diseases, disorders or injuries. American law allow labs to dispense umbilical cord stem cells only to scientists and physicians engaged in FDA approved animal studies and human clinical trials. That means it is only legal to treat a child with its own umbilical blood stem cells when the treatment conditions are FDA approved (Payne 2005). Mexico allows a much wider use of umbilical blood stem cells. That causes Dr. Jensen, a trained radiologist to move from Illinois to Mexico to learn the umbilical stem cell technology. He believes this method is to be one of the body's ablest tools for self-repair. He founded his own very modern Dr. Jensen Stem Cell Clinic in San Luis, Mexico. The technology devices of the clinic let his patients feel like being in a doctor's office in the United States. His staff speaks English. Dr. Jensen assures on his webpage, Umbilical Stem Cells, that over 16,000 procedures with umbilical stem cells have so far been performed without any negative after effects.

Parents of newborns face a critical decision as to whether to

What could be more important?
Children's (Downtown Seattle)

collect and save their baby's umbilical cord blood. Neither the child nor mothers are at risk when harvesting umbilical cord blood. A newborn baby has a volume of 180 ml. umbilical cord blood. Cord blood is rich in hematopoietic stem cells, which can be stored in private cord blood banks for personal future use. It can also be donated to the public. The cost of storing is $2000 for collection and approximately $120 annually.

For donated cord blood, no collection fee may be charged. The first sibling-donor cord blood transplant was performed in 1988. Dr. Pablo Rubinstein developed the first cord blood program at the New York Blood Center. He is able to provide matched cord blood stem cells to patients.

The private use of cord blood was controversially viewed. Chances of an autologous transplant (using your own cord blood stem cells) are rare (Nietfled et al. 2008). Thus, ethicists oppose private cord blood banks. The European Union Group of Ethics stated: "Private cord blood banks sell a service and make a promise they cannot deliver. The activities of such banks raise serious ethical criticisms" (Opinion of the European Group on Ethics 2004).

In comparison, human embryonic stem cell research has solely undeliverable promises, but ethic committees seldom complain. Although human embryonic stem cells were thought to offer potential cures, there is no cure at present. However, in 2008 the first child was successfully treated by autologous cord blood stem cells (Hayani et al. 2007). Only five in 100,000 children can profit from their own umbilical cells.

Usually the use of autologous cord blood cells for the treatment of childhood leukemia is contra-indicated. Autologous cord blood carries the same genetic defects as the patients themselves. Some researchers think it is not effective for the treatment of genetic diseases. At the same time, scientists promise the possibility of somatic nucleus transfer for therapeutic cloning to create embryos for use in biomedical research. It has to be considered that donated somatic cells also carry the same genetic defect of the patient they want to treat. Based on the latest scientific and clinical evidence the successful use of autologous cord blood transplants was described for pediatric cancer neuroblastoma, aplastic anemia, Juvenile Type 1 Diabetes (Haller et al. 2008) and cerebral palsy (Duke University, Phase I clinical trial).

It is easier to match cord blood transplant with "non-related" patients. Public umbilical cord blood banks have greater therapeutic benefits. Even though a disadvantage is that donated umbilical cord blood will provoke immunological reactions thirty percent of the time. The volume and quantity of the cells in a typical cord blood collection is only enough to treat a sixty-pound patient. For adults more than one donation is necessary for therapeutic purposes. The optimum transplant dose can also be reached when cells are grown in the lab prior to use. In 2008, a study from the University of South Florida showed that umbilical cord stem cells slow down Alzheimer's progression in mice. Low-dose infusion of cord blood cells into Alzheimer mimicking lab-rodents could reduce two markers of the disease (Phan and Post 2008).

In the last few years, subpopulations isolated from cord blood have been shown to differentiate into neural-like cells when administered to "patients" with neurodegenerative diseases, brain ischemia or spinal cord injuries. The study was performed in animals, clinical trials are under way (Arien-Zakay et al. 2010).

More and more sources of stem cells that can renew tissue supply are discovered, for example in menstrual blood. Researchers found that menstrual stem cells could turn into many different tissue types, including bone, blood vessel, fat, brain, lung, liver, pancreas and heart cells (Mitchell 2007).

11 The empty womb
11.1 Affordable health care begins with breast-feeding

Hans Urs von Balthasar a Swiss theologian stated that each thing has two perspectives. You can look at it as a fact or as a mystery. If you view humans as a fact, they are random products at the edge of the cosmos. Humans do not depend on an accidental evolutionary play that simply throws them into the world. Human's existence is surrounded by mystery. Each human being has a purpose and a relationship with the environment (Beisner et al. 2008). Theologians even go further when proposing that each human is a thought of God. Thus, behind each of us is an idea and plan (Ratzinger 2000). Bernhard d'Espagnat, a leading quantum physicist, believes that science alone cannot fully explain the "ultimate reality", the nature of a being. "Mystery is not something negative that has to be eliminated," he said. "On the contrary, it is one of the constitutive elements of being; those who believe in a spiritual dimension of existence and life are fully right" (d'Espagnat 2009). On October 28, 2010, Pope Benedict XVI declared that the task of science: "remains a patient yet passionate search for the truth about the cosmos, about nature and about the constitution of the human being. Scientists do not create the world," the Holy Father pointed out. "They learn about it and attempt to imitate it, following the laws and intelligibility that nature manifests to us." He added, "The scientist's experience as a human being is therefore that of perceiving a constant, a law, a logos that he has not created but that he has instead observed. In fact it leads us to admit the existence of an all-powerful Reason, which is other than that of man, and which sustains the world. This is the meeting point between the natural sciences and religion. As a result, science becomes a place of dialogue, a meeting between man and nature and, potentially, even between man and his Creator" (Zenit 2010).

Assume that a young woman-scientist would provoke God with the petition to create her own human being. God would agree but under the conditions that the young woman would have to bring her own created clay. Sad to say, that humans are not even able to create breast milk. Premature infants are only able to tolerate breast milk whereas formulas carry a risk of developing infections or bowel diseases, allergies and vomiting. "Babies were born to be breast-fed," says Director Larraine Lockhart-Borman of the Mothers' Milk Bank of Denver, Colorado. She is engaged in developing programs for breast-milk donations. It is not a question of healthy nutrition and a mother's choice; it may be a question of life or death. Each ounce of donated breast-milk yields $ 4. With that the milk is more expensive than formula. However, the potential benefits are worth it. A milk bank will pasteurize the donated milk. Milk banks did lose their attractiveness with the outbreak of HIV. The seven milk banks in North America accept donations from healthy lactating mothers who undergo medical screening including comprehensive blood tests (Associated Press 2008). It would be extremely practical and beneficial if science would know the pathway to produce human milk. It sounds very simple. Only science does not even know how to make cow-milk out of grass. Could it be that nobody ever thought about that? Because researchers are more occupied in creating more important things like organs instead of wasting their expertise on basics like milk.

Every cow can produce milk. It is not her personal contribution, because humans increase milk production by adequate feeding and breeding. Actually the dual purpose, the combined production of milk and meat let humans manipulate cattle animal feed. A cow would be fine eating only grass, but humans determined it to be unprofitable. Already in 2002 a third of the world cereal production was fed to cows, stated the Food and Agriculture Organization (of the United Nations) FAO. Increasingly, farmland is cultivated for animal feed. In the former Soviet Union rising meat consumption created economic problems. In 1990, Soviet livestock were eating three times as much grain as Soviet citizens. The production of human food is stepping back, whereas a cow could really eat what humankind is unable to eat. It is a fact that a cow can produce milk from grass but it is still a mystery for researchers how the cow does it. To manipulate cattle fodder seems not to be the smartest decision as we have also learned from Mad Cow Disease.

The digestive system of ruminating mammals is designed to digest green fodder. After softening of the grass in the first stomach, Rumen, the partly digested mass, also known as cud, is regurgitated and chewed again. Feeding grain to such ruminating animals will result in a decreased digestive efficiency. 25% of grain will go undigested and will be excreted. Grain feeding is not at all nutritious for ruminants. The stock will suffer from increasing lactic acidosis due to the rapid fermentation of grain by the stomach bacteria. Increasing lactic acids can lead to dehydration, heart failure, kidney failure and even death (2011 Global Fodder Solutions/Turkiye).

India is not engaged in beef productions, for reasons we all know. Frances Moore Lappe wrote in her book, Diet for a Small Planet: "For every sixteen pounds of grain and soy fed to beef cattle in the United States, we only get one pound back in meat on our plates. The other fifteen pounds are inaccessible to us, either used by the animal to produce energy or to make some part of its own body that we do not eat (hair, bones)."

Lappe infers that milk production is more efficient. India could never engage in beef production.

Americans live in the hamburger land, occupied by McDonald's restaurants. The hamburger has become the great export to China and Japan. It is difficult for Westerners to watch India's worship of the cow. The protection of the cow has been called a "lunatic obstacle". 200 years of British foreign occupation with all its imperialistic attempts could not change the issue of India's "sacred cow".

The term has come to explain the stubborn loyalty to a longstanding institution, which impedes natural progress. Millions of humans suffer from undernourishment. The progressive Western civilization sees the country as underdeveloped, overpopulated, overcrowded, undereducated, steeped in superstition and the most backward nation in many respects. Our compassionate suggestions are that India should abandon the religion and kill and eat the cow. Westerners like to consider our ways as the best. Indians should get rid of the burden of poor logic by replacing superstitions with rational thinking. We have to know that the average farming household in India has approximately one acre. The weather makes the purchase of machinery useless. Heat and monsoons do not allow a long growing season. India's zebus (known as humped cattle) are ideal as work animals to plow the soil. Thus, no heavy machinery is overly compacting the soil. Most Indians are vegetarian. The cow also supplies milk (Winter 2008).

Even so India's overpopulation of grass-eating cows is mainly causing environmental damage. In some regions, overgrazing is especially caused by "scrub" cattle that are kept simply as manure-makers. The "dung cakes" are widely used for cooking and heating purposes. In the Hindu religion, the cow is sacred. Indian scriptures declare that the cow is a gift from the Gods to the human race. A cow represents the Divine Mother that sustains all human beings and brings them up as her very own offspring. The majority of Indians are Hindus, for most of whom the killing of cattle and eating of beef is unthinkable because this species is the most sacred of all creatures. According to the moral concepts of Hinduism, it is forbidden under death penalty to harm cattle (Sambraus 1989).

American Bison

The moral protection is given to an animal. It is respected even though the population is one of the poorest in the world. Nevertheless, different world-views create a tension between technical and logical possibilities.

The protected animals are roaming freely through Indians cities and causing traffic jams and eating garbage. A microchip should help in small steps. The cow has to undergo a strange procedure while being forced to swallow the chip. With the use of a special scanner, the owners of garbage eating cows can be identified. The benefit of controlling animals is to get better milk. The milk of the freely roaming cows is not pasteurized and covers fifty percent of the milk supply (New Delhi news 2006).

In all big cities in India, cows are loose on the streets to scavenge in garbage. The owner of the cows are not real farmers and leave the animals on the road to fend for themselves and to save the cost of natural feed and grass. The animals only survive on the waste food, which often stack in polythene bags. The plastic cannot be digested and accumulate over time in the cow's first stomach. There is no space left in the rumen to start natural digestion. Calves born from affected cows are small in size. The animals are in constant pain caused by packed rumen plastic. For the first time a cattle hospital in Anantapur was reporting on December 4th, 2010 about street roaming cows fed on garbage who ingested plastic. The hospital team has removed as much as 65kg of plastic from one cow's stomach during a life-saving operation. Veterinarians are speaking about a cruel and immoral practice, the animals look even "fat" and will bring a lot of money if sold. In the end the animals are suffering and are sentenced to a slow and cruel death (Karuna Society 2011).

11.2 Bonum facere, versus the "me too drug"

Hippocrates an ancient Greek physician (460 BC) is the father of medicine. He published the "Corpus Hippocraticum". The most famous part is the Hippocratic oath, a document on ethics, good medical practice and morals. Medical graduates regularly take the oath, which is almost identical to the original version. The oath emphasizes the duty of the physician to be of benefit to all people regardless of status or class (bonum facere). The physician also commits himself to apply dietetic measures for the benefit of the sick according to his ability and judgment; He will keep them from harm and injustice (nihil vocerer). The patient himself has an autonomy to be treated in a well-meaning manner, fairly and not in a harmful way. The medical ethics provides guidelines for the conduct of medical personnel. In this regard, it has to be considered that medical treatment is getting more and more expensive. Consumer surveys always reveal that many relate to the experience that using cheaper materials is doing a shoddy job. This might be logical because nobody can be convinced that a cheaper Chinese car is better than an expensive luxury Cadillac. Dental patients prefer a root canal treatment, followed by a crown, as being better than the much cheaper extraction of the tooth. But is "more" automatically better? On one side nobody would want to swallow drugs when having the conditions of degenerative hip arthritis which causes a person to endure years of limited mobility and pain, when a more expensive hip replacement is available. Patients will demand therapies when they are available. On the other hand, in a lot of instances, certain tests or procedures or medication were or are not reasonably indicated and in such cases more is not better, but even worse. "Many people seem to believe the old adage that you get what you pay for even if that is not always right", argues internal medicine physician Dr. Gaulte in the retired doctor's thoughts. He further points out that governmental agencies are often put in charge and will conduct comparative effectiveness research. The economic situation would therefore have an influence in downplaying the advantages of new therapies. In addition, a government agency might limit funding for research with a conflict of interest (Gaulte 2010).

Health insurance and its financing is one of the never-ending topics in the United States and elsewhere. The rising costs in health care prevent the physician from applying all measures technically possible. Does a health check become more like a car repair? The patient is an object, the disease a "faux pas". We increasingly seem to believe in the promise to repair any physical damage with advanced medical technologies (Reiter 2002, b). Do the technological advances make the patient-physician relationship less human? For instance, we already have robotic medicine in which the machines are intermediaries between the physician and the patients.

Embryos seem to be everywhere, only not where they are naturally supposed to be. Research with embryonic stem cells derived from humans is controversial. The combat begins with the donation of human egg cells. Researchers try to find a way that human egg cells are no longer the limiting factor for stem cell research. Already in 1998 an American pharmaceutical company tried to enucleate egg cells from cows to substitute the core with a human somatic nucleus (Marshall 1998). To allow a better study of stem cells, human genes are inserted with nuclei into egg cells void of its original animal nucleus. In 2003 a Chinese research group proclaimed the successful creation of stem cell lines obtained from cloned human (DNA)-rabbit (egg cell) beings. Meanwhile Great Britain allows generating cytoplasmic admixed creatures (embryos that contain human and animal matter). In the United States, it is still forbidden to clone such creatures (Weiss 2005). Researchers see also a huge advantage in being more successful in elucidation of the pathway of differentiation when they harvest embryonic stem cells in a later stage of development. The law allows an embryo to live until day fourteen. It is questionable whether an embryo can even live longer than fourteen days in a Petri dish. Great Britain allows the grows of cloned embryos as far as the development of the "primitive streak," which means until the anterior-posterior axis is developed and before organ differentiation occurs. After that stage, the embryological process does not show any prominent landmarks. Biomedical researchers agree to extinguish embryos that have reached the stage of the "primitive streak". This convention and boundaries are necessary. Otherwise, researchers would try to keep embryos alive longer until cell differentiation begins (Greely 2006).

Currently the biggest ethical controversy is about the issue of establishing stem cell lines from the reconstituted eggs of cytoplasmatic admixed human-animals.

In some experiments, human embryonic stem cells are inserted into monkey brains. Human embryonic stem cells are foreign to the immune system of the monkey species. In addition, they are not differentiated. In an undifferentiated status, the cells might be responsible for causing tumors. Is the suffering of the animals in these cases justified? In the end, all experiment participants have given their lives, no matter if monkey or human embryos or mice are involved. Animal rights activists rarely object to such experiments. Scientists currently investigate potential therapies with already differentiated human fetal neural stem cells. However, concerns have been raised over the safety of this experimental therapeutic approach.

In February 2009, Russian scientists provided the first example of a donor-derived brain tumor. Human fetal neural stem cells were injected into a boy with ataxia telangiectasia, a rare neurodegenerative inherited disease, which affects many parts of the body and causes severe disability. Four years after the first treatment, the patient was diagnosed with a brain tumor. Further studies showed that the tumor was derived from the donated transplanted fetal neural stem cells. The Moscow research team suggested that more work is urgently needed to assess the safety and efficacy of such therapies (Amariglio et al. 2009). The developed brain tumor was exactly the tumor researchers can already treat with cord blood stem cells. It is interesting, that "differentiated" fetal neural stem cells cause tumors. It has been proposed that with the discovery of the pathway of differentiation the goal of human embryonic stem cell therapy is accomplished. As this example shows, differentiation does not prevent tumor formation.

11.3 Aspirations versus technical realities

Human embryonic stem cells are seen as the gold standard in research in every respect. Embryonic stem cell research funds were never lacking. When federal laws in the USA prevented the support, research was subsidized by private funds. In Germany, stem cell research can only be performed with embryos imported from outside the country. Politicians and scientists are speaking about a double moral standard. Only a few research centers are permitted to work with human embryonic stem cells. High priority research projects permit the import of cells from abroad (Hüsing 2003). One hundred German scientists were asked in the so-called Delphi Study about the future efficiency of adult versus human embryonic stem cell research. The study took place from June 2003 to May 2004. The survey investigated therapeutic, social, political and ethical aspects. In the opinion of the scientists the following are the limiting factors for human embryonic stem cell research: the funding, the national or international demands, the cooperation with foreign research centers, the social acceptance, the trained personnel, the availability and access to research information and the costs of treatment covered by health insurances. The participants of the study expressed their disappointment mainly because they so very much desired a different outcome. They expected to have gained more knowledge on the subject by now. They hoped to have moved the field of stem cell research much further ahead than they actually have. Researchers were finally discouraged as they realized that human embryonic stem cell research takes more than wishing or hoping.

German scientists are much more optimistic about the performance of adult stem cell research. Ninety percent are confident in that field. Forty percent of the researchers fear that human embryonic stem cells will develop tumors. They propose that tumors and stem cells speak the same language. In addition to ethical doubts, serious therapeutic handicaps exist which result in the researchers loss of confidence in future embryonic stem cell therapies. Total skepticism exists about cloning and germ line treatment, a gene therapy consisting in a willful modification of the genetic material in cells of the germ line to manipulate the next generation. The clinical application of human embryonic stem cells therapy is currently still in its infancy. Assuming the therapy becomes possible, it would result in an increase of embryo centered controversies. Scientists clearly demand adult stem cell innovations, as it will be the most important scientific target for the next ten years. The course of biomedical progress is often unpredictable and the settings of research priorities should be made toward placenta, amniotic fluid, and umbilical cord. According to that inquiry, seventy percent of the German researchers are convinced about a non-existing future for human embryonic stem cells. Embryonic stem cell research is risky. Unfortunately, this kind of research was excessively idealized from its very beginning (Wiedemann et al. 2004). The German debate on stem cell research is burdened with the role of human genetic research during the Nazi regime in Germany. Adult stem cells, in contrast to human embryonic stem cells, are free of moral and ethical concerns (Knowles 2004).

The International Stem Cell Organization is urgently warning people who get human embryonic stem cell therapies in foreign countries. It is far too dangerous to approach a treatment. Numerous medical centers world-wide exploit the hopes of patients, even though they will maintain their serious life threatening status and will not get any help. Facilities are far from conventional medical standards, in countries like India, the Philippines, Mexico, Thailand, China, Barbados, Turkey, Costa Rica, Russia and others that offer expensive human embryonic stem cell therapies. Without reliable treatment to protect the patient's safety, the promises of the clinics can never be fulfilled. The international society for human embryonic stem cell research ISSCR is very concerned about the carelessness toward such patients.

12 Request for researchers

12.1 Extracorporeal embryos

Many have applauded new human embryonic stem cell technologies. People who see benefits in human embryonic stem cell research are also confronted by ethical and scientific dilemmas. They believe that stem cell research is a viable and almost magic solution to many diseases. However, there are still many caveats. Ethical problems and questions arise when using embryos created for in vitro fertilization but left unused, or embryos created specifically for research. It is still a long way to approach clinically useful results and implement new biomedical technologies. Can we balance the possible benefit of human embryonic stem cell research inventing new therapies, and the ethical problems, raised by this research?

For what purposes are human embryonic stem cells used in research? Why is science confident that we have found in them a remedy for almost every disease? Why are we still without the necessary therapeutic success after a decade of research in the field and twenty nine years (2010) of experience in animal embryonic stem cell research? All things considered, the question remains as to why it is so problematic to achieve effective therapies? Stem cells are unspecialized and do not have any tissue-specific structure. Researchers rely only on the potential of the cells. It is conceivable that they could give rise to specialized cells like heart cells, muscle cells, blood cells or nerve cells. The four most important areas of its application are:

1) Studies of human embryonic stem cells may yield information about the complex events that occur during human development, in order to understand the genetic and molecular signals that trigger stem cell differentiation. External signals for cell differentiation are also of great interest for researchers. The specific factors and conditions include: chemicals secreted by other cells, physical contact with neighboring cells, and certain molecules in the microenvironment.

2) It took twenty years of trial and error to learn the proper conditions to grow stem cells in cell culture. Pluripotent stem cell cultures hold great promise in assessment of the toxic effects of biological and chemical drugs. The benefit for scientists would be to understand why cancer cells survive despite very aggressive treatments. Furthermore, the numbers of animal studies and human clinical trials might be reduced by the use of human embryonic stem cell cultures.

3) Pluripotent embryonic stem cells might offer a possible source of replacement cells and tissue based-therapies. Scientists want to be able to reliably direct the differentiation with identification of the specific sets of signals, because the need of transplants far exceeds the number of available organs.

4) A fourth reason for studying stem cells is gene therapy. A gene that provides a missing or necessary protein is smuggled into the specific organ for a therapeutic effect. A gene has more chances to be expressed long-term when introduced into

a less rather than more differentiated cell. Scientists would like to create differentiated tissue derived from such a gene introduction. Current gene therapy can solely modify germ lines. Germ lines refer to sperm and egg cells in which the disease can be eradicated and thus it can be prevented from being passed on. Germ line therapy is forbidden world-wide. This technique can be abused for the production of designer babies.

Human embryonic stem cell research is usually hailed as a great improvement for public health and is rarely critically analyzed. The benefits and risks remain unclear. For instance, the announcement that cultured mouse embryonic stem cells can develop into egg cells. This experiment would fix the problems with manipulation of germ lines, infertility treatment and stem cell research (Hübner et al. 2003). It is still unclear whether the cells can be fertilized and develop into normal embryos (Badura-Lotter and Schubert 2008). Once the mouse experiment data can be translated to humans, the donation of human egg cells will be no longer necessary. The desire to manufacture egg and sperm from embryonic stem cells is not completely new.

Previously it has been proposed that eggs and ovarian tissue from abortions can be used as source for egg cells. In 2003 scientists from the University of Pennsylvania describe a technology to provide egg cells. They managed to harvest immature ovaries from mice and coaxed them to develop. The process is repeatable in humans. In 2003 researchers from Israel and Holland announced at the Nineteenth Annual Meeting of the European Society for Human Reproduction and Embryology in Madrid, that they had successfully removed undeveloped egg cells from four month-old aborted fetuses. They theorized that they could further mature the eggs to use them for IVF or human embryonic stem cell research. "The aborted fetus could serve as biological mother of IVF children", said the Israeli gynecologist Dr. Tal Brion-Shental. "I am fully aware of the controversy about this," Dr. Brion-Shental explained to the reporters, "but probably, in some places, it will be ethically acceptable" (The New Atlantis 2003)

Sperm cells have already been generated from embryonic mouse cells (Dennis 2003). There are also common thoughts of scientists to reprogram induced pluripotent stem cells until they are gamete cells. Science hopes to avoid the need to harvest female oocytes. It would ease the pressure on women to donate eggs. There is a danger that in this way their acts of altruistic donation may be demeaned. Procuring ova for non-reproduction purposes is influencing the status of women in society. "Women may become at risk of being alienated from their reproductive labor, and their ova could become at risk of becoming the means to achieve the aims," stated a speaker of the Australian Parliament (Rickard 2005). We might end up with a global trade in human eggs. Third World Countries might be exploited as ovum donors, because many of these countries have no national ethics committees or guidelines (Cregan 2005). When we commercialize human life, we risk turning those embedded social traditions into instrumental matter open to economic speculation.

What cost would we pay, if this happens (Cregan 2005)? At what cost can the new technique develop and who would benefit from it? Whose lives would be made better by stem cell therapies? In case it should become a routine therapy, would the poorer patients, the unemployed without health care have access to any of these technologies? Many researchers consider human embryonic stem cells suitable for unlimited research applications. However, the potential of stem cells is purely speculative and it is premature to decide to prioritize human embryonic stem cells. To find out which cells are going to be suitable for new therapies more research needs to be done with adult, fetal cells from cord blood, induced pluripotent stem cells and so on.

According to presently existing knowledge, significant problems need be solved. Actual cures are many years away, since research has not progressed to the point where even one cure has yet been generated by human embryonic stem cell research. For instance, stained embryonic stem cell cultures for still hypothetically transplantation therapies can become contaminated and then are useless. Furthermore, substantial obstacles are seen in uncontrolled tumor growth, poorly understood differentiation, the undetermined integration of the cells, confusing growth and healing effects and immunological challenges. An enormous technical effort is necessary to overcome these serious difficulties (Badura Lotter and Schubert 2008). Yet human embryonic stem cells remain for many unmatched in their potential for treatment of a wide variety of diseases and health conditions. They are still viewed as the most optimal cells for the greatest advancements in biomedical research. President R. Murphy of the California Institute of regenerative Medicine believes that human embryonic stem cells are the gold standard for regenerative medicine. Harvard University stem cell biologist Kevin Eggan declared in an interview with The Wall Street Journal: "Human embryonic stem cells will be better, even if they are more complicated politically" (Hotz 2008).

Alternatives are at hand. Veterinarians are currently using fat derived stem cells therapy for aging pets. Adult stem cells are known to boost the body's healing capability and repair tissue. Hip dysplasia, arthritis and other joint, ligament and tendon aliments in companion animals like dogs, cats and horses are healed by the pet's own cells. A little of the animal's fat is collected and sent to a veterinary company. There the regenerative stem cells are isolated, loaded into syringes and shipped back to the local veterinarian to be injected into the pets ailing joint. The aging animals receive their own cells, which will prevent rejection problems. Dr. Robert Harman, founder of the Veterinary Stem Cell Company says: "Nearly 300 animals have had the procedure and eighty-five percent of those we have been able to follow up with have experienced slight to very substantial improvement" (Cornwell 2010). Adult stem cells have already been used to successfully cure many diseases in animals, which is only a "side" benefit.

Many believe that insufficient attention has been given to explore the potential of adult stem cells. Their usefulness is very often underestimated, because of the promises attributed to human embryonic stem cell research. Nevertheless, private embryo-research facilities who are increasingly in need of funds still find private donors, because the moral hazard issue is not relevant to them.

Private donors supported human embryonic stem cell research with contributions of more than $190 million (Journal of Nature 2008). In 2008 several states, including Connecticut, Maryland, Illinois and New York funding these research. The California Institute of Regenerative Medicine, received $3 billion in state bond funds. In addition, in 2008 the California Institute of Regenerative Medicine has had awarded $519 for human embryonic stem cell research. $13 million were dedicated to stem cell research which does not involve the destruction of human embryos (Hotz 2008).

In November 2007, stem cell researchers in Japan and the USA, University of Wisconsin, announced a technique avoiding the use of human embryonic stem cells. Independent research teams successfully converted human skin cells to mimic embryonic stem cells. The so-called "induced pluripotent stem cells" were infected with a man-made virus that carries four extra genes designed to reprogram (re-differentiate) the cell's machinery. Even so, both teams urged to continue human embryonic stem cell research, or at least to proceed together, because the therapeutic potential of induced pluripotent stem cells remains undefined. The genetically engineered virus randomly alters a cell's genetic structure, raising the risk of cancer. For human embryonic stem cell researchers it is a consequence that even more studies of human embryonic stem cells are necessary to verify the reliability of the new technique (Susan Solomon, executive director of the New York Stem Cell Foundation). Researchers will continue human embryonic stem-cell research with human embryos, "until the reprogrammed cells can be made safe and stable enough for clinical experiments." The genetically reprogrammed cells cause Harvard University stem-cell biologist Kevin Eggan to worry about their use in animals or people or to model diseases (Hotz 2008). Dr. James Thomson and Dr. Shinya Yamanaka cautioned that they still must confirm that the reprogrammed human skin cells are identical to stem cells they obtain from embryos. While those studies are under way, Dr. Thomson and others believe it is premature to abandon research with stem cells taken from human embryos (Kolata 2007). The reprogramming technique hailed immediately a long stymied solution to a religious and political impasse even though the problems associated with retroviruses and oncogenes used for reprogramming need to be resolved before iPS cells can be considered for human therapy. In vitro reprogramming of somatic cells into a pluripotent stem cell is viewed as very controversial in the scientific community. Stojkovice, stem cell researcher in Spain, doubts the potential given to iPS cells will be superior to human embryonic stem cells. He stated long-term studies would likely show that human embryonic stem cells are irreplaceable. For him researchers are wasting their time, funds and their intellectual potential in conducting such compatible studies (Stojkovice 2008). Alexander Meissner from Harvard-University in Boston observed that unfortunately only a couple of cells can be reprogrammed from more than hundred-thousand attempts (Frankfurter Allgemeine Zeitung 4.4.2009). Epigenetic modifications influence gene expression pattern and provide a unique signature of a cell differentiation status. Without external stimuli or signaling events, the cell identity remains stable. Epigenetic mechanisms regulate all biological processes from conception to death, including genome reprogramming during early embryogenesis and gametogenesis, cell differentiation and maintenance of a committed lineage. Key epigenetic players are DNA methylation. Induced pluripotent stem cells appeared to display more methylation than embryonic stem cells. The existence of an epigenetic-based cellular memory delays or incompletely reprograms the differentiated nucleus to an embryonic state. Methylation supports intrinsic epigenetic memory. In other words, methylation can independently confer cellular memory and enable reprogramming events. Efficient demethylation is likely a key (if not the only) rate-limiting step to improving the efficiency and outcomes of iPS cells. Fresh oocyte factors carry out demethylation and human embryonic stem cells have an advantage over iPS cells (Deng et al. 2009). The role of the epigenetic state of the cells defines an unexpected decisive hurdle for efficient reprogramming of iPS cells.

Davor Solter, a developmental biologist at the institute of Medical Biology (IMB) in Singapore, expects that science will convert iPS cells into sperm and egg cells in the near future. In fact it was already used to help a patient with a gene-defect who was unable to produce his own sperm cells. Renée Reijo-Pera of Stanford-University successfully reprogrammed somatic cells from the patient to fully functional sperm cells

(Frankfurter Allgemeine Zeitung, 4.4.2009). What moral values and rights are given to such created embryos would be unclear (Le Ker 2008).

In June 2008, Hans Schöler from the Max-Plank-Institute for molecular Biomedicine in Münster reported that only two more genes are required to be transported with viral vectors into somatic cells for reprogramming them to pluripotent embryonic-like stem cells. In this way, one can avoid the use of the carcinogenic c-Myc-Gen as vector. This research has been done with a mouse model; however, these results cannot be transferred to humans yet. In July 2008, at a Stem Cell Meeting in Dresden, Germany, Schöler presented the successful isolation of germline (testis) derived pluripotent stem cells (gPS). These cells with their easy and safe production are superior to all previously reprogrammed cell lines. Already in 2006, Gerd Hasenfuss and Wolfgang Engel from the University of Göttingen, Germany, discovered pluripotent stem cells in mouse testes.

12.2 Advanced success with adult stem cells

Professor Herzog at the University of Bonn, Germany pointed out the still remaining difference of 1200 genes between reprogrammed iPS cells and embryonic stem cells. Thus, Herzog wonders about the purpose of trying to imitate human embryonic stem cells. At present, the profound knowledge and vast understanding needed to use embryonic stem cells as therapeutics is still lacking. For Herzog it is very difficult to understand announcements of human embryonic stem cells therapeutics that allure with empty promises. So far, efforts to develop stem cell therapies have only been realized with adult stem cells. They present a tremendous opportunity for a dramatic improvement in healing injuries and in the cure of diseases. Adult stem cells are used successfully without safety complications for a great benefit. The most important advancement is that patients use their own therapeutic cells to address a wide range of medical concerns. For the past forty years, researchers' work has been most efficient in the area of adult stem cells. Therefore, it is an unreasonable argument that science would depend upon the knowledge and experience gained by human embryonic stem cell research. Dr. Herzog is convinced that science did back the wrong horse by starting human embryonic stem cell research (Rehder 2007).

Naturally, the freedom of research plays a major role in the academic and scientific discussion. Unfortunately, promises of potential treatments are of much higher relevance than ever before. Therefore the core of the debate is compromising the protection of the embryo in favor of technical advances even if they are only hypothetical. Embryonic stem cells are viewed to have a greater potential for development and differentiation compared to adult stem cells. Differentiation of cells almost never involves a change in the DNA sequence itself. Kidney or liver cells have only different physical characteristics but the same genome. Differentiation only changes a cell's size, shape, membrane potential, metabolic activity and responsiveness to signals. The changes are largely due to highly controlled modifications in gene expression. During embryogenesis a single fertilized egg cell, the zygote changes into many cell types. During morphogenesis "totipotent" stem cells become pluripotent and finally turn into fully differentiated cells like neurons, muscle cells, epithelium, blood vessels etc. which continue to divide. All of the cells of the organism are genetically homogenous, they contain exact the same DNA. Structurally and functionally they differ which is caused by different gene expression (Breburda et al. 2003). The DNA carries the genetic information and dictates the protein synthesis. The DNA alone can never give rise to an embryo.

Epigentic processes turn on or off the right genes in utero to determine if the cell shall differentiate into a brain or kidney cell. Epigenes change the gene activity, but do not involve alterations to the genetic code. Those changes can be inherited. A recent article in Time (Jan 6, 2010) claimed that: "The new field of epigenetics is showing how your environment and your choices can influence your genetic code-and that of your kids. And poor eating habits of the mother can lead to heart problems in children" (p. 49). According to scientists, "powerful environmental conditions (near death from starvation, for example) can somehow leave an imprint on the genetic material in eggs and sperms" (p. 50) that can affect one's offspring. It is believed that in humans grandchildren of grandparents who gorged themselves die earlier. It is reported that baby lotions containing peanut oil might be partly responsible for peanut allergies. Anxiety during pregnancy may lead to asthma (p.53). Smoking can predispose children to disease and early death (p. 50). Changes in the environment can place marks on the top of the gene. While genes do not change, the epigenes influence them. Researchers explain it like this: "If the gene is the hardware, then the epigene is the software." That is, "you're going to have the same chip in there, the same genome, but different software. And the outcome is a different cell type" (p.51). Recently scientists have begun to realize that epigenetics could explain certain scientific mysteries that traditional genetics never could find (Cloud 2010).

Stem cell research is nothing else than discovering the responsible mechanism for cell differentiation. Scientists have known about epigenetic markers since 1970. Until the late nineties the epigenetic phenomena was not taken seriously as the main event the DNA is playing in embryonic development. Human embryonic stem cell researchers did not consider them playing a part in cell differentiation until now. In general, cell differentiation is also a common process in adult stem cells. They divide and create fully-differentiated daughter cells during tissue repair and during normal cell turnover. You might say adult stem cells are already doing what human embryonic stem cell researches are attempting to find out. Therefore, the long way is to observe embryonic stem cells, or the short cut is to investigate adult stem cells. This explains also the hopes that embryonic stem cell based regenerative medicine could be only applied in a remote distance. As a matter of fact, embryonic cells have a higher failure rate than adult stem cells. It is doubtful that tumor development and false differentiation of transplanted embryonic stem cells can ever be avoided. Therapeutic problems demonstrate many reasons for skepticism. The tumorigenic potential of embryonic stem cells and maybe even induced pluripotent stem cells might limit their usefulness. A focus and advancement for human embryonic stem cells is in the use of the cells in toxicological and pharmacologic cell-culture experiments. However, induced pluripotent stem cells are equal if not superior in regard to these tests (Vogel and Holden 2007).

Despite the current therapeutic limitations of induced pluripotent stem cells, they can be used almost immediately in laboratory experiments to study the progress of genetic diseases and to test drugs. The establishment of efficient methods of the extraction and accumulation of adult stem cells and induced pluripotent stem cells makes an immediate benefit available. In addition researchers have unlimited access to them and they can be used right away as a patient specific resource. Further alternatives to embryonic stem cells are: umbilical cord cells, pluripotent bone marrow stem cells, and amniotic fluid cells. Whereas, stem cells from amniotic fluid might be more-than-multipotent (Hengstschläger 2003). Their widespread use for testing pharmaceuticals will consequently lead to medical success. Especially because these alternative sources of stem cells only require storage for a short time in culture media. Chromosomal, genetic and epigenetic mutations occur in human embryonic stem cell cultures. They are caused by the culture milieu when embryonic stem cells are maintained too long. Main obstacles have to be overcome to find an appropriate way to culture the cells. Human embryonic stem cells require a culture to maintain the undifferentiated state and pluripotency. Nevertheless, the cells undergo spontaneous mutations. Especially long term cell culturing damages the DNA sequences. Similar genomic alterations are commonly observed in human cancer cells which makes the lines unusable for eventually therapeutic purposes. Equally important are the epigenetic roots of cancer. They cause deregulation of human development. Human embryonic stem cells have a unique epigenetic signature. Very little is known about epigenetic changes and variations in the DNA sequences and epigenetic regulation in human embryonic stem cells during long-term culture (Maitra et al. 2005) IPS cells have a clear advantage inasmuch as they will not mutate (Foroni et al. 2007 and Yoon et al 2005).

12.3 Very best alternative

Meanwhile the stem cell debate is interdisciplinary. Research on the controversial issue involves ethical, scientific, political and economic aspects. Scientists face a human egg shortage and argue this is preventing them from medical breakthroughs, even preventing them from proceeding with therapeutic cloning research. The accessibility of oocytes for research poses a serious ethical challenge to society. The need for eggs will cause risks to the women who provide them.

Researchers are searching for alternative sources of human embryonic stem cells.

It is known that embryonic stem cells possess genetic abnormalities when collected from embryos that are not appropriate for in vitro fertilization. It is presumed that discarded embryos lack the developmental potential to become a healthy child. In 2004, William Hurlblut designed a somewhat artificially abnormal embryo when he introduced the altered nuclear transfer. His goal was to invent through genetic engineering and cloning processes an embryo-like entity that cannot implant and thus is unable to develop beyond the blastocyst stage. Hundreds of human oocytes were required for the Hurlblut cloning process. The development stops at the stage in which the engineered genetic defect becomes manifest. Whatever the efforts of Hurblut are, the destruction of disabled embryos, which are not viable in the long term, is ethically unacceptable. With his method he will not even provide healthy stem cells, even though he is wasting hundreds of human oocytes.

Robert Lanza of Advanced Cell Technology in Worcester, Massachusetts pioneered embryo biopsy. The technique, also known as blastomere biopsy, is performed by removal of one or two cells (blastomeres) from certain six to eight cell embryos that undergo pre-implantation screening for particular genetic disorders. At the early stage, every cell is omnipotent and has the same potential for development (Klimanskaya et al. 2006). Couples undergoing in vitro fertilization treatment have one cell removed before the embryo is implanted in the womb. The separated cell could be grown in a blastocyst. The early embryo, the blastocyst, must be destroyed to harvest stem cells. It is suggested that embryonic cells might also be stored in case the child is in need of treatment for future diseases with its own stem cells.

Finally, the identical twin will be brought to term, whereas his sibling is destroyed for research purposes. The argument to avoid the destruction of an embryo is therefore misleading. In addition, evidence is given that embryo-biopsy is not entirely harmless. Studies have shown that embryos subjected to prenatal diagnostic and then transferred to the womb are less likely to result in pregnancy and live birth (See at al. 2007).

During in vitro fertilization occasionally an abnormal embryo can be produced, which is carrying three pre-nuclei, whereas normally the zygote has two pre-nuclei. Researchers from Israel used sixty abnormal embryos and were able to develop nine of them into the blastocyst stage. These were then used for research purposes (Suss-Toby et al. 2004). The scientist justified the use of the abnormal embryos because they were non-viable. Such arguments are disturbing because the in vitro fertilization generated these disabled embryos.

Currently the further utility of stem cells is not given, even if researchers choose to destroy embryos to derive natural pluripotent stem cells or use the artificially derived way to obtain induced pluripotent stem cells. Scientists will have the same hurdles and hopes with embryonic stem cells as with iPS cells, namely figuring out their differentiation.

It is well known that embryonic stem cells can turn malignant. The characteristics of undifferentiated embryonic stem cells are very similar to those of cancer cells (See et al. 2007 and Andrews et al.2005). Semi-differentiated stem cells may also cause cancer. Prior to their use, it would be necessary to investigate the malignant potential of embryonic stem cells. Scientists admit that it is difficult to use them in the near future.

The University of Wisconsin, Madison received on August 4, 2008, together with two other US Universities $ 9 million as research funds to solve the mysteries of differentiation. Prof.

Thomson prefers to work with the stem cell lines he isolated in 1998 (Wahlberg 2008, c). On March 9, 2009, President Barack Obama lifted, by Executive Order, the Bush administration eight-year ban on federal funding on embryonic stem cell research. This move should expand stem cell research, even as it diminishes the use of UW Madison researcher James Thomson's stem cell lines.

The key argument for human embryonic stem cells research is the hope to cure Alzheimer, Parkinson, Diabetes, Cancer and so forth. At the present, is it difficult to substantiate the promises. The technology is rapidly improving. Thus, the alleged advantage of embryonic stem cells should have already been demonstrated. Instead, adult stem cells prove to be the more viable option. Breakthroughs in adult stem cell research are taking place. Human embryonic stem cells have little merit toward adult stem cells. No study of the therapeutic success of embryonic stem cells has been published so far (Stejskal 2008). In addition, immune rejection does not occur in adult stem cells, whereas it will accompany human embryonic stem cell research. Our bodies quickly recognize and try to kill foreign tissues implanted in them. Using cells from oneself is likely to avoid the compatibility problem (Postovit et al. 2007, Andrews et al. 2005).

The only possible benefit of human embryonic stem cells is to use them for germ line therapy. However, worldwide it is strictly forbidden to design humans. Not even the most extreme reason justifies germ line therapy. Germ line therapy may be able to pick out traits and eradicate diseases from future generations. Parents cannot choose what kind of baby they want not even under the rationale of avoiding inherited genetic diseases. Essential genetic alterations made to germ cells, or early stage embryos are transferred to all the future generations. We might induce harmful mutations by that step which we then pass on to our descendants. We are unaware of the multiple functions genes have. A change in one gene certainly may lead to complications and unanticipated interactions. Changing ourselves may create descendants so different from us that "they do not even recognize us as fully human" (Wertz 1998).

Canis culinarius cloned?

Society might not be satisfied and quickly move on to enhancements when once a certain new biotechnology is started. If we know how to add genes, we might think we have the pathway to make better human beings. Our children already begin to be "genetic independent" from us. Are legal, technical and social ramifications sufficient to influence and convince us to carefully use new biomedical knowledge? Can scientists of tomorrow hesitate to employ their knowledge, simply because they feel uncomfortable or have ethical doubts? "The world will be much worse off if parents are allowed to choose their child's trait," declares Katherine Circle, a student of South Dakota State University in 1998. Such a designed child would become a pure object dependent on a strange will and preference of another person. That would be most contradicting to the dignity of each person. No human life should be started or ended as an object. Humans would become instruments of their fellow humankind in order to explore them. Kant is referring in the formula of humanity to the so-called *Instrumentalisierungsverbot:* "To preserve human dignity it is forbidden to see human as objects or to exploit them in order to gain a benefit from his traits". Do we create in modern society human embryos with pure economic rationality for the special purpose to experiment on them and to gain scientific knowledge? Does the *homo oeconomicus* consider the human embryo as reduced to an object to obtain stem cells? Shall we regard the embryo outside of the womb as a person? Do we admit that he has fundamental human rights even though his appearance in such an early developmental stage is not reassembling the so familiar human shape? Are human rights bound to a certain shape? Fundamental rights are given to everyone without preconditions and independent from his appearance or the existence of additional qualities (Schockenhoff 2008).

Are the goals and promises of human embryonic stem cell research justified? Is the issue of human embryonic stem cell research discussed too one-sided or uncritical, by only focusing on a desired medical progress? Is the legitimating of further scientific endeavors completely dependent upon whether researchers succeed in fulfilling the utopian promises to heal? Who will take moral and legal responsibility if the hopes of healing which are raised in patients cannot be fulfilled?

######

References

AAVS American Anti-Vivisection Society. Animal Cloning. Frequently asked Questions about animal cloning. 2. Why do researchers want to clone animals? 2006 http://www.aavs.org/animalcloning_faq.html#faq2

Abisambra JF, Fiorelli T, Padmanabhan J, Neame P, Wefes I, Potter H. LDLR expression and localization are altered in mouse and human cell culture models of Alzheimer's disease. PLoS One. 2010 Jan 1; 5 (1):e 8556.

Ackerman F. The economics of atrazine. Int J Occup Environ Health. 2007 Oct-Dec: 13 (4): 437-45

Adams J. U. Scoping out a new breed of rules. Are genetically engineered fish and meat coming soon? We examine the Food and Drug Administration's regulations. Los Angeles Times, Monday January 26, 2009 latimes.com/health F-F6

Adams R. Government funding for stem cell research blocked by US court. The Guardian.co.uk/world/richatd-adams-blog/2010/24/ Aug, 24. 2010

Adjaye J. Stammzellen Berliner Genforscher warnt vor Chimären-Produktion; James Adjaye: Hybrid-Embryonen "problematisch" Mitteldeutsche Zeitung, 20.05.2008

Alton D. Lord of Liverpool, Health: Stem Cell Therapy Publications & Records 3 March 2009; Column 688

Alton D. Opening Statement to the House of Lords Select Committee on Stem Cell Research- 19th November 2001

Alok J. British IVF pioneer Robert Edwards wins Nobel prize for medicine. Robert Edwards, the British scientist who pioneered IVF, was responsible for the conception of Louise Brown, the world's first test-tube baby. Guardian.Co.UK, Monday October 4 2010

Amariglio N., Hirshberg A., Scheithauer B.W., Cohen Y., Loewenthal R., Trakhtenbrot L., Paz N., Koren-Michowitz M., Waldman D., Leider-Trejo L., Toren A., Constantini S., Gideon Rechavi G. Donor-Derived Brain Tumor Following Neural Stem Cell Transplantation in an Ataxia Telangiectasia Patient. PloS Medicine, Vol 6 Issue 2 e1000029, p 1-11, Feb. 17, 2009

1st Annual Wisconsin Stem Cell Symposium: Neural Stem Cells, BioPharmaceutical Technology Center Institute (BTCI) 5445 E. Cheryl Parkway, Madison (Promega Campus), 19. April 2006

Andrews PW., Matin M.M., Bahrami A.R., Damjanov I., Gokhale P., Draper J.S. "Embryonic Stem (ES) Cells and Embryonal Carcinoma (EC) Cells: Opposite Sides of the Same Coin," Biochemical Society Transactions33.6 (December 2005): 1526–1530. PMID: 16246161

Anissimov M. What is Parthenogenesis? WiseGeek 09 September 2010

Antonucci F., Rossi C., Gianfranceschi L., Rossetto O., Caleo M. Long-distance retrograde effects of botulinum neurotoxin A. J. Neurosci. 2008 Apr. 2:28(14):3689-96.

PMID: 18385327 [PubMed - indexed for MEDLINE

ARD, Tilmann J. Menschen auf Eis gelegt (HR) Kryofrostung als Segen und Horrorvision 23. July 2008, 22:45 Uhr im Ersten

Arien-Zakay H., Lecht S., Nagler A., Lazarovici P. Human umbilical cord blood stem cells: Rational for use as a neuroprotectant in ischemic brain disease. Int J Mol Sci. 2010; 11 (9): 3513-3528

Aristotle, Nicomachean, Ethics, Politics. In the Works of Aristotle Translated into English, edited by W. Ross, 12 vols., Oxford, Clarendon Press, 1921-1952

(C), Politics book I chapter 2, 1253a 2

Asbell B. The Pill A Biography of the Drug that changed the world. Random House. ISBN: 0-679-43555-7, chapter one, The Conception, 1995

Associated Press Mother donates breast milk. The milk went to a milk bank that supplies programs like that of a state hospital that uses if for premature babies. Wisconsin State Journal, Dec. 8. 2008, page: A3

Aziz A., Soucie E., Sarrzain S., Sieweke MH. MafB/c-Maf deficiency enables self-renewal of differentiated functional macrophages. Science, 2009 Nov 6; 326 (5954): 867-71

Badura-Lotter G. und Schubert L.: Stammzelle: Was können wir wollen. Mensch Medizin, Gen-ethisches Netzwerk Januar 2008

Backer M., Monkey stem cells cloned. Nature 447, 891 (21 June 2007) nature.com/nature/journal/v447/n7147/full/447891a.html

Baier A., USF studies show link among Alzheimer's disease, Down syndrome and atherosclerosis. USF Health News Alzheimer's and Neurosciences, Research Really Matters, January 14, 2010

Barncard C. Scientists, policymakers call for stem cell funding. University of Wisconsin news/ 18361, September 7, 2010

Batalion N., 50 Harmful effects of genetically modified (GM) foods. RAW-WISDOM .com http://www.raw-wisdom.com/50harmful. 2009

Beck R., EU hat Bedenken gegen Fleisch und Milch geklonter Tiere. www.nikidesaintphalle.de/?EU+hat+Bedenken+gegen+Fleisch+und+Milch+geklonter+Tiere 17.1.2008

Beisner E. C., Cromartie M., Derr T. S., Hill P. J., Knippers D., Terell T. Environmental stewardship in the Judeo-Christian Tradition. Jewish, Catholic, and Protestant wisdom on the environment. Acton Institute 2008, ISSN 1-880595-15-x p: 7

Berner Zeitung 413. US-Genforscher Craig Venter warnt vor Klonversuchen bei Menschen (Berner Zeitung Nn: 2.05.2001) selected

gene-tech messages from the world Swiss & German press
unifr.ch/biochem/BIOTECH/BIO-401-450.html#413

Bethge P. Germany's Mystery Cow Disease Causing Calves to Bleed to
Death. Spiegel, New Magazine: Environment, Farmers, Health
03/27/2009

Bhutani N., Brady J.J., Damian M., Sacco A., Corbel S.Y., Blau H.M.
Reprogramming towards pluripotency requires AID-dependent DNA
demethylation. Nature. 2010 Feb 25;463 (7284):1042-7.

Bienengentechnik
www.bienengentechnik.de/fix/docs/files/PM%20M%Fcllverbrennun
%20Honig%20080924.pdf 2008

Billette de Villemeur T., Gelot A., Deslys J.P., Dormont D., Duyckaerts C.,
Jardin L., Denni J., Robain O. Iatrogenic Creutzfeldt-Jakob disease in three
growth hormone recipients: a neuropathological study. Neuropathol Appl
Neurobiol. 1994 Apr; 20(2): 111-7

Bintz K. L., International Sport-horse Registry, 939 Merchandise Mart,
Chicago, IL 60654, Feb. 1995

Bio-Genica, Genetic Engineering and Manufacturing, Inc. 2004-2008,
Genpets™ Patented Biotch, November 08, 2008, Genpets.com

Byers D. Octuplets' mother, who already has six children, turned down
selective abortion. Times Online UK, January 30, 2009 timesonline.co.uk,
article5618449.ece

Bogdahn U. Reperatur des Nervensystems. Neues aus Wissenschaft und
Welt. Science ORF, 2000

Bömelburg H. Mysteriöse Krankheit, Hilfe für den Baummenschen, Stern,
Wissenschaft, Medizin 22.11.2007 URL

Bond L. "The Surviving Sibling," Nat'l RTL News, Sept. 25, 1986

Bonfranchi I. Menschen mit Trisomie 21 sterben aus. Soziale Medizin,
(1996). 1, 38 – 39.

Brazelton T.R., Rossi M.V., Keshet G.I., Blau H.M. From Marrow to Brain:
Expression of Neuronal Phenotypes in Adult Mice Science, 1 December,
2000:Vol. 290, 5497, pp.

Golos T.G., Bondarenko G.I., Dambaeva S.V., **Breburda** E.E., Durning M.
On the role of placental Major Histocompatibility Complex and decidual
leukocytes in implantation and pregnancy success using non-human
primate models. Int J Dev Biol. 2010 54 (2-3): 431-43,

Dambaeva S.V., **Breburda** E.E., Durning M., Garthwaite M., Golos T.G.
Characterization of decidual leukocyte populations in cynomolgus and
vervet monkeys, Journal of Reproductive Immunology J Reprod. Immunol.
2009 Jun; 80(1-2): 57-69. Epub 2009 Apr 26 (b)

Drenzek J.G., **Breburda** E.E., Burleigh D.W., Bondarenko G.I., Grendell
R.L., Golos T.G. Expression of indoleamine 2,3-dioxygenase in the rhesus

monkey and common marmoset. J Reprod Immunol. 2008 Jul; 78(2):125-33. PMID: 18490060 [PubMed]

Bondarenko G.I., Burleigh D.W., Durning M., **Breburda** E.E., Grendell R.L., Golos T.G. Passive Immunization against the MHC Class I Molecule Mamu-AG Disrupts Rhesus Placental Development and Endometrial Responses. J Immunol. 2007 Dec 15; 179(12):8042-50.PMID: 18056344 [PubMed - indexed for MEDLINE]

Golos T.G., Bondarenko G.I., **Breburda** E.E., Dambaeva S.V., Durning M., Slukvin I.I. Immune and trophoblast cells at the rhesus monkey maternal-fetal interface. Methods Mol Med. book chapter, 2006;122: 93-108. PMID: 16511977 [PubMed - indexed for MEDLINE]

Breburda E.E., Dambaeva S.V., Golos T.G. Selective Distribution and Pregnancy-Specific Expression of DC-SIGN at the Maternal-Fetal Interface in the Rhesus Macaque: DC-SIGN is a Putative Marker of the Recognition of Pregnancy. Placenta 2006, 27, 11-21 PMID: 16310033 [PubMed - indexed for MEDLINE]

Slukvin I.I., **Breburda** E.E., Golos T.G. Dynamic changes in primate endometrial leukocyte populations: differential distribution of macrophages and natural killer cells at the rhesus monkey implantation site and in early pregnancy. Placenta. 2004 Apr; 25(4):297-307. PMID: 15028422 [PubMed - indexed for MEDLINE]

Kaiser M.E., Merrill R.A., Stein A.C., **Breburda** E., Clagett-Dame M. Vitamin A deficiency in the late gastrula stage rat embryo results in a one to two vertebral anteriorization that extends throughout the axial skeleton. Dev Biol. 2003 May 1; 257(1): 14-29. PMID: 12710954 [PubMed - indexed for MEDLINE]

Breburda E., Wirth T., Leiser R., Griss P. The influence of intermittent external dynamic pressure and tension forces on the healing of an epiphyseal fracture. Arch Orthop Trauma Surg. 2001 Sep;121(8):443-9. PMID: 11550830

Breburda E., Wirth T., Leiser R., Griss P. Zellproliferation nach externer dynamischer Fixierung bei Tibiaepiphysenfugenverletzungen beim Lamm als Modell. (Cell proliferation after dynamic external fixation of the tibial growth plate in a sheep model), pp 328-333; in: Verletzungen von Becken bis Fuss im Kindesalter, book chapter Editors: S. Hofmann v.Kapherr, S. Berger, O. Beck, Achen, Shaker Verlag, 2001 ISBN 3-8265-8676-X

Breburda E., and Schnettler R. Dynamische externe Fixierung verbessert Heilung von Läsionen der Wachstumsfuge. Osteosynthese International, Johann Ambrosius Barth Verlag Heidelberg (2000) 8: 4-8 (Dynamic external fixation improves healing of lesions of the growth plate)

Breburda E. Auswirkungen mechanischer Einflußgrößen auf das Längenwachstum nach Unterbrechung der Gefäßversorgung der proximalen Tibiaepiphysenfuge beim Lamm als Modell, Dissertation, Giessen, 1996 (Effect of mechanical factors on longitudinal growth of the growth plate of the proximal tibia after interruption of vascular supply in a sheep model, Ph.D. thesis)

Breburda J. (scientific advisor) Desert Problems and Desertification in Central Asia: The Researchers of the Central-Asia-Desert-Institute in Ashchabad. Berlin, Heidelberg and New York, Springer, 1999

Breburda, J. (1983): Bodenerosion und Bodenerhaltung, 128 Seiten, 42 Abbildungen, DLG-Verlag Frankfurt, 1983

Broley C. Plight of the American bald eagle. Audubon Magazine Vol. 60, 1958; pp. 162-171

Brons, I.G., Smithers L.E., Trotter M.W., Rugg-Gunn P., Sun B., de Sousa C., Lopes S.M., Howlett S.K., Clarkson A., Ahrlund-Richter L., Pedersen R.A., Vallier L. Derivation of pluripotent epiblast cells from mammalian embryos. Nature 2007, Jul 12: 448, (7150): 191–195

Brown J.L. StarLink Biotechnology Food & Agriculture. The Pennsylvania State University 1/3/2003 http://biotech.cas.psu.edu/articles/starlink.htm

Butler R.N. Scientists appeal to president-elect Obama, Column sent to the president-elect, Guest column, Wisconsin State Journal, Friday Nov. 21, 2008 page A12

Callier M. Artificial spider silk could improve body armor, parachutes. The official web site of the U.S. Air Force 2/28/2008

Caplan A. New IVF dilemmas make old fears seem quaint. Twins for a 70-year-old? Louise Brown's doctors didn't envision that. Health- Health care-Breaking Bioethics-msnbc.com 7/24/2008

Campbell D. Women will be paid to donate eggs for science. The Observer, Sunday Feb. 18, 2007

Cann R. L., Stoneking M. and Wilson A.C. "Mitochondrial DNA and Human Evolution" Nature 325 (1. Januar 1987) Seite 31-36

Carroll S.B. Hybrid's may thrive where parents fear to tread. New York Times , D2 September 14, 2010

Castiello U, Becchio C, Zoia S, Nelini C, Sartori L, Blason L, D'Ottavio G, Bulgheroni M, Gallese V. Wired to be social: the ontogeny of human interaction. PLoS One 2010 Oct 7; 5 (10):e13199

Cho M. and Magnus D. Issues in oocyte donation for stem cell research. Science (2005) 308:1747–1748

Chomsky N. Language and Mind (New York, Harcourt, Brace, Jovanovich) 1972 page 67

Cibelli J.B., Grant K.A., Chapman K.B., Cunniff K., Worst T., Green H.L., Walker S.J., Gutin P.H., Vilner L., Tabar V., Dominko T., Kane J., Wettstein P.J., Lanza R.P., Studer L., Vrana K.E., West M.D. Parthenogenetic Stem Cells in Nonhuman Primates: Science 295 (2002) 819.

Cloud J. Why your DNA isn't your destiny. Time Health and Science Jan 6, 2010

Coghlan A. Hybrid hearts could solve transplant shortage. NewScientist Health, Magazine issue 2711, 03 June 2009

Cohen A. Everyman's Talmud: The Major Teaching of the Rabbinic Sages (New York, Schocken Books, 1975 reprint of 1949 edition), Seite 39-40, 68

Collinge J., Sidle K., Medas J., Ironside J. and Hill A. Molecular analysis of prion strain variation and the etiology of 'new variant' CJD. (1996)

Connor S. Watchdog allows embryo selection for donor tissue. The Independent Science Thursday, 13 Dexember 2001

Cornwell L. Stem cell research offers new options for ailing dogs, cats and horses. Health and Wellness, Alternative Medicine, Published October 01, 2010

Cregan K. Ethical and social issues of embryonic stem cell technology. Intern Med J 2005;35:126-7.

Creutzfeldt H.G. Über eine eigenartige herdförmige Erkrankung des Zentralnervensystems. Z. Ges. Neurol. Psychiat. 1920. 57: p. 1-20.

Crii-Gen: Seralini G-E., Cellier D., Spiroux de Vendomois J. Committee for Independent Research and Information on Genetic Engineering Report on NK 603 GM corn produced by Monsanto company June 2007

Croxatto H.B., Diaz S., et al. Plasma progesterone levels during longterm treatment with levonorgestrel silastic implants. Acta Endocrinologica. 1982; 101: 307-311.

Crozet and Lehmann Prions: where do we stand 20 years after the appearance of bovine spongiform encephalopathy? Med. Sci. (Paris). 2007 Dec; 23(12): 1148-57

Crystal D. The Cambridge Encyclopedia of Language, 2nd ed. (New York: Cambridge University Press, 2002) p.230

Dalai Lama, The Compassionate Life, Tenzin Gyatso, the fourteenth Dalai Lama, Wisdom Publications, Boston 2003

Dealler S. Post-exposure prophylaxis after accidental prion inoculation. Lancet. 1998 Feb. 21; 351 (9102): 600. PMID: 9492809

D'Espagnat B., Die Realität ist nicht in den Dingen, FAZ.NET 3/16/2009

de Fried E.P., Ross P., Zang G., Divita A., Cunniff K., Denaday F., Salamone D., Kiessling A., Cibelli J. Human parthenogenetic blastocysts derived from noninseminated cryopreserved human oocytes. Fertil Steril. 2008 Apr; 89(4):943-7. Epub. 2007 Aug 13.

PMID: 17706204 [Pub Med - indexed for MEDLINE]

Deng J., Shoemaker R., Xie B., Gore A., Leproust E.M., Antosiewicz-Bourget J., Egli D., Maherali N., Park I.H., Yu J., Daley G.Q., Eggan K., Hochedlinger K., Thomson J., Wang W., Gao Y., Zhang K. Targeted bisulfite sequencing reveals changes in DNA methylation associated with nuclear reprogramming Nat Biotechnol. 2009 Mar 29 Epub ahead of print

Dennis C. Developmental biology: Synthetic sex cells. Nature. 2003 Jul 24; 424 (6947): 364-6. PMID: 12879036

Develin K. Baby selected to be free from breast cancer gene. The first Baby genetically screened to be free from a potentially deadly breast cancer gene. Telegraph.co.uk, Friday 09 January 2009

Devitt T. Embryonic stem cell culturing grows art to science. Wisconsin University of Wisconsin-Madison, News/18664, Nov. 14,2010

Dewan S. DNA tests to reveal if possible record-size boar is a pig in a poke. New York Times, Saturday, March 19, 2005

DFG   Stammzellforschung-deutschland-lang-0610.pdf   Aktuelles. Stellungnahmen, 2006

Seite 6

Dor J., Lerner-Geva L., Rabinovici J., Chetrit A., Levarn D., Lunenfled B., Maschiach S., Modan B. Cancer incidence in a cohort of infertile woman who underwent in vitro fertilization fertile Steril 2002: 77:324-327 a) Seite 152

Doherty A.S., Mann M.R., Tremblay K.D., Bartolomei M.S., Schultz R.M. Differential effects of culture on imprinted H19 expression in the preimplantation mouse embryo. Biol. Reprod. 2000; 62: 1526-1535.

Draper J. et al. Recurrent gain of chromosomes 17q and 12 in cultured human embryonic stem cells. In: Nature Biotechnology, BD 22, Nr. 1, Januar 2004

Eilperin J. Female Sharks Can Reproduce Alone, Researchers Find. Washington Post, Wednesday, May 23, 2007

Ehmann R. Verhütungsmittel – verhängnisvolle Nebenwirkungen, über die man nicht spricht, In: Empfängnisverhütung. Fakten, Hintergründe, Zusammenhänge, Hrsg. v. SÜSZMUTH, Roland, Holzgerlingen 2000, 109-271 a= 152 b= 236

Engdahl F.W. Seeds of Destruction. The Hidden Agenda of Genetic Manipulation. Global Research, 2007 ISBN 978-0-937147-2-2 (Reviewed von Stephen Lendman 22. Jan. 08)

Enzensberger H. M. Die Fatalität des Denkens. Kafkas Sätze. Frankfurter Allgemeine Zeitung 11. Juli 2008

Ertelt S. Company claims first patient treated with embryonic stem cells, not the case. LifeNews.com, October 11, 2010

Ethics Committee of the American Society for Reproductive Medicine. Financial incentives in recruitment of oocyte donors. Fertile Sterile 2004; 82:Suppl 1: S. 240-S. 244.

Ettinger R. The Prospect of Immortality. Ria University Press ISBN/ASIN: 097434723X, ISBN-13: 9780974347233, Plus Additional Comments 1964-2005

European Patent Office (EPO), No European patent for WARF/Thomson stem cell application. Nov 27 2008, www.epo.org/topics/news/2008/20081127_de.html

Fagin D. and Lavelle M. Toxic Deception: How the Chemical Industry Manipulates Science, Bends the Law and Endangers your Health. Common Courage Press, ISBN 978-1-567511628, 2 edition July 2002

Fagothey Austin, Right and Reasons, Ethics in theory and practice, based on the teaching of Aristoteles and St. Thomas Aquinas 1958 a) page 284; b) page 286-7; c) page 206; d) page 208; e) page 256-257;

Feller B., Neegaard L. Obama to lift federal stem-cell funding limits. Policy will have immediate impact on researchers. Wisconsin State Journal, Saturday March 7, 2009 p A1, A6

Flader J. Submission on the Research Involving Embryos Bill, sub 1414.doc, 13. September 2002

Foote R. H. The historical or artificial insemination: Selected notes and notables, Department of Animal Science, Coronel University, Ithaca, NY 14853-4801, 2001

Foroni C., Galli R., Cipelletti B., Caumo A., Alberti S., Fiocco R., Vescovi A. Resilience to transformation and inherent genetic and functional stability of adult neural stem cells ex vivo. Cancer Res. 2007 Apr 15; 67(8):3725-33. PMID: 17440085

Frankl V. E. Der Wille zum Sinn, S 109 Gaudium et Spes Vgl. 2. Vatikanisches Konzil, Pastoralkonstitution, Nr. 50 und 51

Gajdusek D.C., Zigas V. Degenerative disease of the central nervous system in New Guinea. New England Journal of Medicine, 1957. 257: p. 974-978.

Geisler N.L. Epigenetics Offers New Solution to Some Long-Standing Theological Problems: Inherited Sin, Christ's Sinlessness, and Generational Curses Can be Explained. 2010 http://www.normangeisler.net/epigenetics.html

George P. R. and Cohen E. The President politicizes Stem-Cell Research. Taxpayers have a right to be left out of it. The Wall Street Journal, Tuesday, March 10 2009. Page A13

Geron S. Study finds that insulin-producing beta cells can be reborn. Different pancreatic cells change into beta cells, suggesting new therapy for type 1 diabetes. April 4, 2010 HealthDay, US News and World Report

Garfield E. Carl Djerassi: Chemist and entrepreneur. Chemtech. Page: 534-538, September 1983

Gaulte J. Sometimes more expensive care is better treatment. In Diagnosis and Treatment/ blog www.kevinmd.com 2010/07

Gilbert K., Former top IVF doc: I was 'absolutely horrified' when I realized what I was doing. LifeSiteNews.com, Mon Jun 13, 2011

Gilbertson M., Kubiak T. und Ludwig G. Tox. Environ. Health 1991 33, 455-520, 1995

Goethe von J.W. Dr. Faust in Faust I "Da steh ich nun ich armer Tor"

Goethe von J.W. Dr. Faust in Faust I "Wer fertig ist, dem ist nichts recht zu machen; Ein Werdender wird immer dankbar sein".

Goethe von J.W. 'Was immer Du tun kannst' Dr. Faust II (Schlusssatz)

Graf R. Klonen: Prüfstein für die ethischen Prinzipien zum Schutz der Menschenwürde Begründet von Josef Georg Ziegler - Herausgegeben von Clemens Breuer Moraltheologische Studien Neue Folge (MSNF), Bd. 5 2003, S., 9 ISBN 3-8306-7170-9

Granic A., Padmanabhan J., Norden M., and Potter H. Alzheimer Ab Peptide Induces Chromosome Mis-segregation and Aneuploidy, including Trisomy 21; Requirement for Tau and APP. *Molecular Biology of the Cell.* 2010 Feb 15; 21 (4): 511-20

Greber B., Wu G., Bernemann C., Joo J.Y., Han D.W., Ko K., Tapia N., Sabour D., Sterneckert J., Tesar P., Schöler H.R. Conserved and Divergent Roles of FGF Signaling in Mouse Epiblast Stem Cells and Human Embryonic Stem Cells. Cell Stem Cell. 2010 Mar 5;6 (3): 215-226.

Greely H.T. Moving Human Embryonic Stem Cells from Legislature to Lab: Remaining Legal and Ethical Questions. PLoS (2006) Med 3 (5): e143

Greenwood J.C., Don't be afraid of Frankenfish. The Wall Street Journal, A 23, Thursday September 23, 2010

Greger M. Mad cow disease: Threat of an epidemic. Green left: Australian news http://www.greenleft.org.au/node/11664, Wednesday June 19 1996

Haberlandt E. Ludwig Haberlandt- A pioneer in hormonal contracetption Wien Klein Wochenschr. (2009) 121: 746-749 DOI 10.1007/s0058-009-1280-x, Printed in Austria Springer-Verlag 2009

Habermas J. Technik und Wissenschaft als Ideologie, Frankfurt 1968, S. 155

Hacker J.H. Vizepräsident der Deutsche Forschungsgemeinschaft in Stichtagsregelung behindert deutsche Stammzellforschung, Journal Med. Gesundheitspolitik 07.04.2008

Hayani A., Lampeter E., Viswanatha D., Morgan D., Salvi S.N. First report of autologous cord blood transplantation in the treatment of a child with leukemia. Pediatrics 2007 May; 119(1): e 296-300 PMID: 17200253 [PubMed - indexed for MEDLINE]

Hall L. Live-in families to help anorexia patients. The Sydney Morning Herald smh.com.au http://www.smh.com.au/lifestyle/wellbeing/livein-families-to-help-anorexia-patients-20090907-fees.html September 8, 2009.

Haller M.J. et al. Autologous umbilical cord blood infusion for type 1 diabetes. Exp. Hematol. 36, 710-715 (2008)

Haumer R.M. Empfängnisregelung und ihre bleibende Brisanz. Eine ethische Betrachtung unter Berücksichtigung demographischer Entwicklungen sowie der Notwendigkeit einer guten Sexualerziehung, Diplomarbeit, Philosophisch-Theologische Hochschule St. Pölten, Mai 2008; a) S.39, b) S. 31

He Y., Wu J., Dressman D.C., Iacobuzio-Donahue C., Markowitz S.D., Velculescu V.E., Diaz L.A., Kinzler K.W., Vogelstein B., Papadopoulos N. Heteroplasmic mitochondrial DNA mutations in normal and tumor cells. Nature. 2010 Mar 25; 464(7288): 610-4

Hengstschläger M., Prusa A.R., Marton E., Rosner M., Bernaschek G. Oct-4-expressing cells in human amniotic fluid: a new source for stem cell research? Human Reproduction (2003): (18) Nr. 7, S. 1489-1493, PMID: 12832377

Herbert C.A., Trigg T.E. Applications of GnRH in the control and management in female animals. Animal Reproduction Science Volume 88, Issues 1-2, August 2005, Pages 141-153

Hill J. Scientists successfully create human-bear-pig chimera (meanbearpig) Think Gene a bio blog about genetics, genomics and biotechnology. April 1, 2008

Ho M.W. and Burcher S. GM maize and dead cows. Institute of Science in Society science society sustainability. ISIS Report 13/01/04

Horkheimer M. und Adorno T.W Dialektik der Aufklärung. Philosophische Fragmente, Fischer : Frankfurt am Main 1988, S. 62

Hornbergs-Schwetzel S. Blickpunkt Forschungsklonen, Stand: April 2008, DRZE-Research Cloning.

Hotz R.L. Stem-Cell researchers claim embryo labs are still a necessity. Science Journal, The Wal Street Journal, January 4 2008, Page B1

Hudson, K., Baruch S. and Javitt G. Genetic Testing of Human Embryos: Ethical Challenges and Policy Choices. In Expanding Horizons in Bioethics, ed. Arthur Galston and Christiana Peppard, 2005, 103-122. Dordrecht: Springer.

Huffstutter P.J. Farms downsize with miniature cows. With feed prices up, ranchers see the advantages of smaller breeds of bovines. Los Angeles Times. May 24, 2009

Hulme D. Sechs Dominante Theorien Wirklich nichts Absolutes? VISION ETHISCHE UND NEUE HORIZINTE Theorie 5: Positivismus 2003 http://www.visionjournal.de/03-2/6dom4.htm

Humanae Vitae (Latin „Of Human Life") Encyclical Letter written by Pope Paul VI on the „Regulation of Birth" and the traditional teaching of the

Catholic Church regarding abortion, contraception and other issues pertaining to human life. 25 July 1968

Hügel B, und Süszmuth R. Kommen hormonale Kontrazeptiva als bedenkliche Umweltverschmutzer in Betracht?, In: Empfängnisverhütung. Fakten, Hintergründe, Zusammenhänge, Hrsg. v. SÜSZMUTH, Roland, Holzgerlingen 2000, 503-527, S 504

Hübner K., Fuhrmann G., Christenson L.K., Kehler J., Reinbold R., De La Fuente R., Wood J., Strauss J.F. 3rd, Boiani M., Schöler H.R. Derivation of Oocytes from mouse embryonic stem cells. Science 2003 May 23; 300 (5623), S. 1251-1256. PMID

Hüsing B., Engles E.M., Frietsch R. Menschliche Stammzellen (Human Stem Cells). Study of the Center for Technology Assessment. (TA 44/2003, Bern, Switzerland 2003)

ISIS Report, Independent Scientists an endangered Species. Institute of Science in Society. www.i-sis.org, September 4, 2001

Jalonick M.C., FDA to consider approval of modified salmon. Wisconsin State Journal, Monday, September 20, 2010

James D., Noggle S.A., Swigut T., Brivanlou A.H. Contribution of human embryonic stem cells to mouse blastocystes. Developmental Biology 295 (2006) 90-102

Jastrow R. "Message from Professor Robert Jastrow"; LeaderU.com; 2002

Jia D., Jurkowska R.Z., Zhang X., Jeltsch A., Cheng X. Structure of Dnmt3aL suggests a model for de novo DNA methylation. Nature, 2007 Sep 13; 449 (7159):148-9.

Jiang Y., Jahagirdar B.N., Reinhardt R.L., Schwartz R.E., Keene C.D., Ortiz-Gonzalez X.R., Reyes M., Lenvik T., Lund T. Blackstad M, Du J, Aldrich S, Lisberg A., Low W.C., Largaespada D.A. & Verfaillie C.M. Pluripotency of mesenchymal stem cells derived from adult marrow. Nature 418, 41-49 (4 July 2002) | doi:10.1038/nature00870;

Johannes XXIII., Enz. Mater et Magistra: AAS 53 (1961), S. 447

Johannes Paul II, (Papst von 1978-2005) in Warkulwiz V.P. The Doctrine of Genesis 1-1, universe book 2007, ISBN:978-595-45243-9, page 280-281

Johannes Paul II, Peace with God the creator, Peace with all of creation. Message of his Holiness Pope John Paul II for the Celebration of the World day of Peace, 1. January 1990, published in: Osservatore Romano, Sintesi della catechesi in lingua tedesca, Jan. 26. 2001

Johnson T. U.S. apologizes for giving Guatemalans syphilis. Wisconsin State Journal A10, Saturday, October 2, 2010

Julius Kühn-Institut. Gebeiztes Saatgut als Bienentod, Forscher legen neue Analyse-Ergebnisse vor. Scine XX Das Wissensmagazin, Springer 11. 6. 2008

Jung, Carl Gustav (875-1961), ein Schweizer Mediziner und Psychologe und der Begründer der Analytischen Psychologie.

Jung C. G. Psychologische Typen. Zürich: Rascher 1921; 9. Aufl. als Bd. 6 der Gesammelten Werke, 1960; 17. Aufl. 1994;

Karberg S. Ergebnisse der Forscher-Umfrage, Erfahrungen mit Stammzellen, von 14 Forschern. Politik Kölner Stadt-Anzeiger Feb. 2. 2008

Karuna Society for Animals and Nature. Karuna takes a lead in anti-plastic bag campaign for the animals. 2011 www.karunacociety.org/?page_id=180

Kass L. Beyond Therapy: Biotechnology and the Pursuit of Happiness. Regan Books 2003

Kaiser J. Appeals Court Stays Stem Cell Injunction. Science Insider (American Association for the Advancement of Science) September 9, 2010

KBS Disgraced Hwang Tried to Clone Mammoth, Tiger, Inside Korea, Science and Technology July 26, 2006,

KBS Hwang Admits wrongdoing for Thesis, Domestic, Korean radio, July 4, 2006

Kolata G. Scientist bypass need for Embryo to get stem cells. Science/ the New York Times, 21 November 2007

Kaspar E. Medikamente im Boden und im Wasser. Welche Wirkung haben die nachgewiesenen Substanzen? 25 Mai 2005, Neue Züricher Zeitung

Keder L. The genetics of Prader-Willi Syndrome: An explanation for the rest of us. Chromosome 15 an explanation. Published in Prader-Willi-Syndrome Association. Updated July 2004

Kidd K. Effects of a Synthetic Estrogen on Aquatic Populations: a Whole Ecosystem Study, 1994

Kim D., Kim C-H., Moon J-II., Chung Y-G., Han B-S., Ko S., Yang E., Cha K. Y., Lanza R. and Kim K-S. Generation of Human Induced Pluripotent Stem Cells by Direct Delivery of Reprogramming Proteins. Cell Stem Cell, Volume 4, Issue 6, 472-476, 28 May 2009

Klimanskaya I., Chung Y., Becker S., Lu S.J., Lanza R. "Human Embryonic Stem Cell Lines Derived from Single Blastomeres," Nature 444.7118 (November 23, 2006): 481–485. PMID: 16929302

Kiessling L.L., Klim J.R., Li L., Wrighton P.J. A defined glycosaminoglycan-binding substratum for human pluripotent stem cells. Nat Methods, 2010 Nov 14

Knoepp W. Stem cell therapy: The forbidden therapy that actually works. Exclusive Content Marketing Community, Expert on Healthcare issues 2009-09-24

Knowles LP. A regulatory patchwork--human ES cell research oversight. Nat Biotechnol. 2004 Feb;22(2):157-63. PMID: 14755285

Koroljow D. Vorkommen und Wirkung von östrogen aktiven Substanzen im Futter von Schweinen. Dissertation zur Erlangung des Grades eines Doktors der Veterinärmedizin durch die Tierärztliche Hochschule Hannover Mai 2007

Körtner U. Forschungsethik und Menschenbild, Zum Leitbild medizinischer Forschung. Streitfall Embryonenforschung 2001

Krause Diane S. Multipotent Human Cells Expand Indefinitely: Blood 98 (2001) 2595.

Kreeft P. Finding black and white in a world of grays. Making choices. Practical wisdom for everyday moral decisions. Servant Publications 1990 ISBN 0-89283-638-5, Seite: 127

Ladstätter D. Neue Klone, (Facts, 20.03.2003), Thu, 24 Apr 2003. Biotech Media News, #314

La Merrill M.A. Estrogen in birth control diminishes sex organs in male rats. Environmental Health News. Published by Environmental Sciences. Jan 15, 2010

Latsch G. Are GM Crops Killing Bees, Der Spiegel, 22. March 2007

Le Ker H. Zukunft der Reproduktionsmedizin, 17. Juli 2008, Spiegel Online

Lenzken S.C., Romeo V., Zolezzi F., Cordero F., Lamorte G., Bonanno D., Biancolini D., Cozzolino M., Pesaresi M.G., Maracchioni A., Sanges R., Achsel T., Carrì M.T., Calogero R.A., Barabino S.M. Mutant SOD1 and mitochondrial damage alter expression and splicing of genes controlling neuritogenesis in models of neurodegeneration. Hum Mutat. 2010 Nov 30.

Lerner-Geva L., Geva E., Lessing J.B., Cherit A., Modan B., Amit A. The possible association between in vitro fertilization treatments and cancer development. Int J Gynol. Cancer 2003; 13:23-27

Lieberman B. Couple caught up in debate over fate of frozen embryos. Signon San Diego. Union-Tribune May 25, 2005

LN US–FREIGABE Geklonte Tiere kommen auf die Fleischtheke, Die Welt, 15. Jan. 2008

Lo B., Zettler P., Cedars M.I., Gates E., Kriegstein A.R., et al. A new era in the ethics of human embryonic stem cell research. Stem Cells (2005) 23:1454–1459DOI

Long J. Pregsure BVD still safe despite withdraw in Germany. Farmers weekly interactive. Marketing the Farming Connection. Friday 07, May 2010.

Lord Alton, The Government should back alternatives to embryonic stem cells. Comprehensive coverage of Britain's Conservative Party Jan. 6 2008 platform/lord_alton

Losey J.E., and Vaughan M. The economic value of ecological services provided by insects. Bioscience 56:311-323. 2006

Lubbadeh J. Wissenschaftler machen Zell-Reprogrammierung noch einfacher. Genforschung, Der Spiegel, Wissenschaft 30.06.2008

MacKenzie D. Vets may have spread mad cow disease. New Scientist 14.8.1999, S.24

Magahern J. Obama Futurama, Phoenix Magazine, Lifestyle, Valley News, April, 2009, Page 68

Magnis C. Nutz' die Dinger, bevor sie in den Gully kommen, Cicero Magazin für Politische Kultur, 6.6.2008 www.cicero.de/97.

Magnus D., Cho MK. Issues in oocyte donation for stem cell research. Science 2005; 308:1747-1748.

Maitra A. et al. Genomic alterations in cultured embryonic stem cells. Nat. Genet. 2005 Oct: 37 (10): 1099-103 16142235 Cit:157

Malone et al. In vivo response of honey bee midgut proteases to tow protease inhibitors from potato, Journal of Insect Physiology, vol 44. no 2, 1998, pp 141-7

Malzahn Christian http://www.seelenfluegel.net/erw.html

Marshall E. Claim of human–cow embryo greeted with skepticism. Science 1998 288:1390–1391

Mason E. Preimplantation Genetic Diagnosis (PGD) should be allowed in Germany: study reveals demand for a change in the law. Innovations report, Forum für Wissenschaft, Industrie und Wirtschaft. 6. 28. 2004

Mathews E., Barden T., Williams C.S., Williams J. W., Bolden-Tiller O., Goyal H.O., Mal-Development of the Penis and Loss of Fertility in Male Rats Treated Neonatally with Female Contraceptive 17-Ethinyl Estradiol: A Dose-Response Study and a Comparative Study with a Known Estrogenic Teratogen Diethylstilbestrol. Toxicological Sciences 2009 112(2): 331-343;

Max-Planck-Gesellschaft Dexeptive Model: Stem Cells of Humans and Mice Differ More Strongly Than Suspected (with editorial adaptations by Science Daily staff). Science Daily March 9, 2010

Mayor S. House of Lord's supports human embryonic stem cell research. BMJ medical publoication of the year (BMJ 2001; 322:189 (27 January)

Meadows Donella 1972 Systemverhalten der Erde als Wirtschaftsraum im Zeitraum bis zum Jahr 2100, S. 17

Meadows Donella, Randers J., Meadows Dennis, Limits to Growth, 2004

Meissner A. and Jaenisch R, Generation of Nuclear Transfer-Derived Pluripotent ES Cells from Cloned Cdx2-Deficient Blastocysts, *Nature* 439.7073 (January 12, 2006)

Metclafe C. Water Pollution leads to mixes sex fish. Dec. 6, 2001, Ichthyology in the News. flmnh.ufl.edu/fish/innews/MixedSex2001.html

Mitchell C.B. The Good News and the Bad News about creating embryos for research. Published on The Center for Bioethics and Human Dignity, July 10, 2001

Mitchell S. Menstrual blood tapped as source of stem cells. Finding may offer happy medium between embryonic and other adult cells. MSNBC, Health, cloning and stem cells. Nov 30, 2007. msnbc.msn.com/id/21996417/

Mlynek J. Präsident der Hermann von Helmholtz-Gemeinschaft Deutscher Forschungszentren in: Stammzellen: Stichtag-Verschiebung von 1. Januar 2002 auf 1. Mai 2007, Journal Med Gesundheitspolitik 11.04.2008

Monet D. Plant diversity and seed saving are the foundations of agricultural sustainability. 2009, http: //hubpages.com/hub/ GeneticallyModifiedCropsThreatenTraditionalFarming

Monschein M. Menschliche Stammzellen ohne tierische Produkte geschaffen. Innovations report, Forum für Wissenschaft, Industrie und Wirtschaft 17.03.2005

Moreno J.D., Weiss R., Edwards E., Science Next: Innovation for the Common Good from the Center for American Progress. ISBN-13: 978-1934137185, Bellevue Literary Press April 1, 2009

Möller P. Berlin, Philolex, www.philolex.de/philolex.htm

Muasher S.J., Oehninger S., Simonetti S., Matta J., Ellis L.M., Liu H-C., Jones G.S., Rosenwaks Z. The value of basal and/or stimulated serum gonadotropin levels in prediction of stimulation response and in vitro fertilization outcome. Fertile Sterile 1988; 50: 298-307

Muldrew K. Cryogilogoy – A Short Course. A web-based textbook introduction to the science of cryobiology—the study of living things at low temperatures. Document updated, Mar.23, 1999 http://people.ucalgary.ca/~kmuldrew/ cryo_course/cryo_chap15_1.ht

Müller T. Gen-Lex und Stammzellen: Verzerrte Wahrnehmungen, (Basler Zeitung, 14.03.2003), Thu, 24. Apr 2003

Müller P., Bulnheim U., Diener A., Lüthen F., Teller M., Klinkenberg E.D., Neumann H.G., Nebe B., Liebold A., Steinhoff G., Rychly J. Calcium phosphate surfaces promote estrogenic differentiation of mesenchymal stem cells. J Cell Mol Med. 2008 Jan Feb; 12(1): 281-91.

Mülle-Schmidt R., Auf dem Weg zur Eugenik von unten? Die moralischen Herausforderunge der Vererbungsforschung speisen sich aus ihrer Geschichte, aber auch aus ihren therapeutischen Perspektiven. In Berlin geht heute der Weltkongress zu Ende. Frankfurter Allgemeine FAZ. NET July 18. 2008

Muotri A.R., Nakashima K., Toni N., Sandler V.M., Gage F.H. Development of functional human embryonic stem cell-derived neurons in mouse brain. Proc. Natl. Acad. Sci. USA December 20, 2005 vol. 102 no. 51 18644-18648

Ney P. "Relationship Between Abortion & Child Abuse," Canada Jour. Psychiatry, vol. 24, 1979, pp. 610-620

Nietfeld J.J., Pasquini M.C., Logan B.R., Verter F., Horowitz M.M. Lifetime probabilities of hematopoietic stem cell transplantation in the U.S. Biology of Blood and Marrow Transplantation. 2008;14: 316-322

Norsigian J. Egg Donation Dangers, Gen watch, Volume 18 Number 5 September-October 2005, gene-watch.org/genewatch/articles/18-5Norsigian.html

Norsigian J. "Risks to women in embryo cloning," Boston Globe, February 25, 2005,

O'Brien B. Donor conception a minefield of unregulated practice. The Irish Times, Saturday, June 12, 2010

O'Brien J. Op-Ed: Stem-Cell Research can Promote Life, Dignity and Discovery. Stem cell news. CAMR coalition for the advancement of medical research, June 13, 2008 Orlando Sentinel

Ogbogu U. New law may be the only way out for researchers affected by US stem cell injunction. Stem Cell Network, blog, August 26, 2010

Opinion of the European Group on Ethics, in science and new technologies to the European commission. Opinion number 19. 16th March 2004.

Ott C.H., Matthiesen T.S, Goh S-K., Black L.D., Kren S.M., Netoff T.I., Taylor D.A. Perfusion-decellularized matrix: using nature's platform to engineer a bioartificial heart. Nature Medicine 14, 213-221 (2008)

Pacholczyk T. Fire in the clinic. Making Sense: Bioethics. Catholic Herald, Diocese of Madison Dec. 4th 2008, p 13

Phan K. and Post C. Umbilical cord stem cells slow down Alzheimer's progression in mice. Sun, Mar. 30 2008. christianpost.com/article/20080330/umbilical-cord-stem-cells-slow-down-alzheimer-s progression-in-mice.htm

Papst Benedikt der XVI. Zum Abschluss des Weltjugendtags in Sydney. 20. July 2008

Papst Johannes Paul II Vgl. seine Katechesen in den Generalaudienzen 1979-1984, hrsg. und eingeleitet von Norbert und Renate Martin in zwei Bänden; Johannes Paul II., Die menschliche Liebe und Die Erlösung des Leibes, Vallendar 1985

Papst Johannes Paul II, Apostolisches Schreiben Familiaris Consortio an die Bischöfe, die Priester und die Gläubigen der ganzen Kirche über die Aufgaben der christlichen Familie in der Welt von heute. Zweiter Teil, Ehe und Familie im Plane Gottes Der Mensch, Abbild des liebenden Gottes. 22. November 1981

Park A. The Quest Resumes. After eight years of political ostracism, stem-cell scientist like Harvard's Douglas Melton are coming back into the light-

and making discoveries that may soon bring lifesaving breakthroughs. Time Vol.173, No.5 p.38-43, Feb. 9, 2009

Parlby G. Health Editor, POSITIVE NEWS, UK (Okt. 99)

Patel T. France reels at latest medical scandal, New Scientist magazine, issue 1884 of 31 July 1993, page 4

Paul-Ehrlich-Institute (Agency of the German Federal Ministry of Health) The Committee for Veterinary Medicine Products, CVMP at the European Medicine Agency (EMA) follows a request by the Paul-Ehrlich-Institute to "Blood Sweating of Calves". Friday, July 16, 2010

Payne A. G. The Steenblock Research Institute Handbook on Umbilical Cord Stem Cell Therapy (Dr. Fernando Ramirez's Program in Mexico), Anthony G. Payne, 2005

Pennings G., de Wert G., Shenfield F., Cohen J., Tarlatzis B., Devroey P. ESHRE Task Force on Ethics and Law 12: oocyte donation for non-reproductive purposes. Hum. Reprod. 2007 May;22 (5): 1210-3. Epub 2007 Mar 8.PMID: 17347168

Pitsch M. Obama stem-cell bump would benefit area, UW and Dane country will see a windfall if federal money flows after the new president ends funding ban. Wisconsin State Journal, Nov.19. 2008

Pius XII., Address to Italian Catholic Union of Midwives, 29. October 1951: AAS 43 (1951), p. 846.

Pluhar W.S. (translated) Immanuel Kant, Critique of Judgment, Hackett Publishing Co., 1987, ISBN 0-87220-025-6, p 531-532

Pollmer U., Warmuth S. BSE: Wurde der Rinderwahnsinn durch infiziertes Fleischmehl verursacht? Lexikon der populären Ernährungsirrtümer (2000) Eichborn Vlg., Ffm.: ISBN: 3821816155

Postovit L. M., Costa F. F., Bischof J. M., Seftor E. A., Wen B., Seftor R. E., Feinberg A. P., Soares M. B., Hendrix M. J. "The Commonality of Plasticity Underlying Multipotent

Tumor Cells and Embryonic Stem Cells," Journal of Cellular Biochemistry 101.4 (July 1,

2007): 908–917 PMID: 17177292

Postrado L. Zebra gives birth to horses. Manila Bulletin. August 16, 2010

Priehn–Küpper S. Klonen von Säugetieren, Dollys Zoo-das kopierte Leben Zahnärztliche Mitteilung zum 97, Nr. 12, 16.06.2007, page 28-32

PSRAST World renowned scientist lost his job when he warned about GE food- The Pusztai case. Physicians and Scientists for Responsible Application of Science and Technology. A Global Network, 2000, http://www.psrast.org/pusztai.htm

Purdey M. Ecosystems supporting clusters of sporadic TSEs demonstrate excesses of the radical-generating divalent cation manganese and deficiencies of antioxidant cofactors Cu, Se, Fe, Zn. Does a foreign cation substitution at prion protein's Cu domain initiate TSE? Med. Hypotheses, 2000. 54 (2): p. 278-306.

QuarterHorseNews: Clone Update, The Whole Story. On line. On Time. On Target. http://quarterhorsenews.com/index.php/news/industry-news/69-clone-update-the-whole-story.html 2010

Ramde D. Johnson weighs in on stem cells. Wisconsin State Journal, A5 Saturday, October 2, 2010

Ramsey Colloquium, The Inhuman use of Human Beings. A Statement on Embryo Research by the Ramsey Colloquium. Sponsored by the Institute on Religion and public life named after Paul Ramsey, 1913-1988, First Things 13 July 2003

Randerson J., Aprad Pusztai: Biological divide. The scientist at the center of a storm over GM foods 10 years ago tells James Randerson he is unrepantant. Guradian.co.uk Thursday 15 Januar 2008

Ratzinger J. (Papst Benedikt 16.) Gott und die Welt, Glauben und Leben in unserer Zeit, Ein Gespräch mit Peter Seewald. Friedrich Pustet Verlag Regensburg 2000, page 65 ISBN 3-412-05428-2

Reefhuis J., Honein M.A., Scheive L.A., Correa A., Hobbs C.A., Rasmussen S.A. The National Birth Defects Prevention Study. Assisted reproductive technology and major structural birth defects in the United States. Hum Reprod. 2008 Nov. 16

Rehder S. Stammzell-Debatte 2: Zellbiologe Volker Herzog hält Aufregung für widersinnig. Embryonale Stammzellen sind das "falsche Pferd". Die Tagespost Würzburg 28. November 2007.

Reiter J. Bioethik, Mitschrift vom Wintersemester 2002/2003 von Anke Heinz (a) page 30 Ansprache Pius XII 1944), (b) page 10

Richard JW. Environmental stewardship in the Judeo-Christian Tradition. Jewish, Catholic, and Portestant wisdom on the environment. Acton Institute 2008, ISSN 1-880595-15-x, Introduction p:2-5

Richard et al., 2005, Environ. Health Perspect. 113, 6, 716-720 & Benachour et al., Committee for Independent Research and Information on Genetic 2007, Arch. Environ. Contam. Toxicol. 53, 126–133.

Rickard M. Current issues brief No. 5, 2002-03: Key ethical issues in embryonic stem cell research. Department of the Parliamentary Library, Australia, 2002 [cited 2005 Jan 12]. Available from: URL: http://www.aph.gov.au/library/pubs/ CIB/2002 03/03cib05.pdf

Ritter M. "Frankenfood" is already here. Wisconsin State Journal, A9 Thursday, September 23, 2010

Rosenthal B. E. Natural Health and Longevity Resource Center. http://www.all-natural.com/wildyam.html

Rossing M.A., Daling J.R., Weiss N.S., Moore D.E. Self S.G. Ovarian tumors in a cohort of infertile woman. N Engl J Med, 1994; 331: 771-776

Rötzer F. Britische Forscher haben bereits 270 Mensch-Tier-Embryonen erzeugt, Telepolis 24.6.2008

Rötzer F. Briten billigen die Forschung mit embryonalen Stammzellen, Telepolis 23.01.2001

Sambraus H. H. Atlas der Nutztierrassen, Eugen Ulmer GmbH und Co. ISBN 3-8001-7213-5. S 33

Sherwell P. LA delivers first designer-baby clinic. World Today, Telegraph New York March 2 2009

Schneider R. U. Kleiner Mann - was nun? Darf eine Frau ein behindertes Kind abtreiben? Ja, sagt der Soziologe und Bioethiker Tom Shakespeare, obwohl er selbst dann vielleicht gar nicht geboren worden wäre. Kindermacher, NZZ Folio 06/02

Schockenhoff E. Ethische Probleme der Stammzellforschung. Humboldt Forum Recht. Die juristische Internet-Zeitschrift an der Humboldt Universität zu Berlin. 6-2008

Schulte von Drach M. C. Wann kommt das Designer-Baby, Spezial, Frage der Woche, Süddeutsche Zeitung vom 7.21.2008 süddeutsche.de/wissen/603/302599/text/

Schwartz L.B., et al. The embryo versus endometrium controversy revisited as it relates to predicting pregnancy outcome in in-vitro fertilization-embryo transfer cycles. Hum Reprod. 1997; 12: 45-50.

See S. Mastenbroeck et al., "In Vitro Fertilization with Preimplantation Genetic Screening," New England Journal of Medicine 357.1 (July 5, 2007): 9–17; and B. Goldman, "Reproductive Medicine: The First Cut," Nature 445.7127 (February 1, 2007): 479–480.

See L. M. Postovit et al., "The Commonality of Plasticity Underlying Multipotent Tumor Cells and Embryonic Stem Cells," Journal of Cellular Biochemistry 101.4 (July 1, 2007): 908–917

Seidel und Kördel, Publikationen des Umweltbundesamtes Bewertung des Vorkomens und der Auswirkung von infektiösen Biomolekülen in Böden unter besonderer Berücksichtigung ihrer Persistenz, Forschungsprojekt. Fraunhofer-Institut für Molekularbiologie und Angewandte Ökologie, September 2007, a) Seite 16, b) 24/25

Seidel G. E. Superovulation and Embryo Transfer in Cattle. Science, Vol. 211, 23 Jan. 1981, p 353.

Selinka H. C. TSE-ERREGER (PRIONEN) IM BODEN: VORKOMMEN UND INFEKTIONSRISIKO Ergebnisse von Untersuchungen des Fraunhofer Instituts für das Umweltbundesamt 2008 Mai, a Seite 6; b) Seite 4

Seoul National University's report on Dr. Hwang Woo Suk the South Korean researcher who claimed to have cloned human cells. Text of the Report of Dr. Hwang Suk. The New York Times, Science section, January 9, 2006

Seralini G.E. Controversial effects on health reported after subchronic toxicity test: a confidential rat 90 day feeding study. Report on MON 863 GM corn pruduced by Monsanto Company June 2005

Sicro R.A. Environmental stewardship in the Judeo-Christian Tradition. Jewish, Catholic, and Protestant wisdom on the environment. Acton Institute 2008, ISSN 1-880595-15-x, Foreword

Siddiqui A.A., Stanley C.S., Skelly P.J., Berk S.L. A cDNA encoding a nuclear hormone receptor of the steroid/thyroid hormone-receptor superfamily from the human parasitic nematode Strongyloides stercoralis. Parasitol Res 2000 Jul;86(7):613. PMID: 10669132

Smith W.J. First human embryonic stem cell treatment commences. First Things, Monday, October 11, 2010

Smith W. J. Another Cloning „Breakthrough" The world's first phony stem cells. The Weekly Standard, December 31, 2005

Spaemann R. in: über den Beginn der Menschenwürde – Eine Vorlesung am Rande des Katholikentages, Von Jürgen Liminski Die Person beginnt im Augenblick der Zeugung DieTagespost/Germania,27.05.2008

Spaemann R. Sie schlechte Lehre vom guten Zweck. Der korrumpierende Kalkül hinter der Schein-Debatte. Frankfurter Allgemeine Zeitung vom 23.10. 1999

Spicuzza M. Attorney General will not represent state in fight against stem cell ruling. Wisconsin State Journal 9/10/2010

Sputtek A. Cryoperservation of Red Blood Cells and Platelets. Methods in Molecular Biology, Vol. 368., Jun. 5. 2007 page: 283-301

Stein R., and Hsu S.S., NIH cannot found embryonic stem cell reserach, judge rules. The Washington Post, August 24, 2010

Steinbock B. Payment for egg donation and surrogacy. Mt Sinai J Med 2004; 71

Steinbrecher R.A. Ecological consequences of Genetic engineering. In Redesigning Life? By Tokar B. Witwatersrand University Press, Johannesburg, Zed Books, London New York, a = p86, b = p88, c = p97

Steinbrook R. Egg Donation and Human Embryonic Stem-Cell Research. The New England Journal of Medicine Volume: Number 4, 354:324-326, January 26, 2006

Steinhauser M.L. and Lee R.T., Cardiovascular Regeneration: Pushing and Pulling on Progenitors. Cell Stem Cell Volume 4, Issue 4, 277-278, 3 April 2009

Stejskal J. Was sind Stammzellen? Imabe-Info 3/08: Stammzellen, IMABE-Institut, Wien

www.imabe.org

St George-Hyslop P.H. Molecular genetics of Alzheimer's disease. Biol Psychiatry. 2000 Feb 1; 47 (3):183-99

Stojkovic M., Phinney D. Reprogramming battle Egg vs. Virus. STEM CELLS 2008; 26:1-2

Stone R. Gift im Korn, Gefahr durch Arsen. Süddeutsche Zeitung 16. July 2008

Strauer B.E., Yousef M. and Schannwell S.M. The acute and long-term effects of intracoronary Stem cell Transplantation in 191 patients with chronic heart failure: the STAR-heart study. European Journal of Heart Failure, 2010, Vol. 12, Issue 7 pp 721-729

Strauer B.E. in: Nutz' die Dinger, bevor sie in den Gully kommen von Constantin Magnis Cicero Magazin für Politische Kultur, www.cicero.de/97

Supreme Court, Diamond v. Chakrabarty, 447 U.S 303 (1980) June 16, 1980 http://www.bustpatents.com/chak.htm

Suss-Toby E., Gerecht-Nir S., Amit M., Manor D., Itskovitz-Eldor J. Derivation of a diploid human embryonic stem cell line from a mononuclear zygote. Hum Reprod. 2004 Mar;19(3):670-5. Epub 2004 Jan 29. PMID: 14998969 [PubMed - indexed for MEDLINE]

Suttner B. Politik die aufgeht. Stopp, Mr. Wilmuth! Regensburg, den 14.03.05

Süddeutsche.de "Als würden sie Blut schwitzen" Die Symptome erinneren an Ebola: Kälber verbluten innerhalb von Stunden, Die Tiermediziner stehen vor einem Rätsel. Hanno Charisius, 5. März 2009

Szabo L., Down syndrome patients could unlock mysteries of aging. USA Today, 3/22/2010

Talarico G., Piscopo P., Gasparini M., Salati E., Pignatelli M., Pietracupa S., Malvezzi-Campeggi L., Crestini A, Boschi S., Lenzi G.L., Confaloni A., Bruno G. The London APP mutation (Val717Ile) associated with early shifting abilities and behavioral changes in two Italian families with early-onset Alzheimer's disease. Dement Geriatr Cogn Disord. 2010 Jul;29(6):484-90. Epub 2010 Jun 3.

Taylor D.M., Inactivation of SE agents. Br Med Bull, 1993. 49: p. 810-821.

Taylor D.M., Fraser H., McConnell I., Brown D.A., Brown, K.L., Lamza K.A., Smith G.R. Decontamination studies with the agents of bovine spongiform encephalopathy and scrapie. Arch Virol., 1994. 139: p. 313-326.

Technische Universität München. Natürliche Abwehrkräfte machen Maispflanzen fitter. Genetische Grundlage des natürlichen Abwehrmechanismus aufgedeckt. ScineXX Das Wissensmagazin, Springer 23.06.2008

Theim R. Frozen in time: How life insurance can make you immortal. Insure.com The 31 resource for insurance. June 28, 2010

The New Atlantis, My Mother, the Embryo. IVF's Latest: She-Males, Fetal Eggs, and Children of the Unborn. The Editors of the New Atlantis, Number 2, Summer 2003, pp. 94-96.

The New York Times, Alderson Reporting Company. Transcripts of oral arguments before court on abortion case. The New York Times. April 27, 1989: B12.

Thomson J.A., Kalishman J., Golos T.G., Durning M., Harris C.P., Hearn J.P., Pluripotent Cell Lines Derived from Common Marmoset (Callithrix Jacchus) Blastocysts: Biol Reprod 55 (1996) 254-259.

Thomson J.A., Itskovitz-Eldor J., Shapiro S.S., Waknitz M.A., Swiergiel J.J., Marshall V.S., Jones J.M. Embryonic stem cell lines derived from human blastocysts. Science. 1998; 282: 1145-1147. PMID: 9804556

Thorel F., Népote V., Avril I., Kohno K., Desgraz R., Chera S., Herrera P.L., Conversion of adult pancreatic alpha-cells to beta-cells after extreme beta-cell loss. Nature. 2010 Apr 22; 464 (7292): 1132-3

Trujillo Lopez, The truth and meaning of human sexuality. Guidelines for education within the family, Issued by the Pontifical Council for the Family. Vatican City, November 21, 1995

Turley J. The ultimate motion to Sever: Arizona court to decide who gets head of dead woman. Bizarre Science Society Published: February 28 2010

University of Pittsburgh Schools of the Health Sciences (2007, April 17). What's In The Water? Estrogen-like Chemicals Found In Fish

Venter C. in: US-Genforscher Craig Venter warnt vor Klonversuchen bei Menschen, Berner Zeitung/Nu/2.05.1 Seite 64 Rubrik Vermischtes, Titel Gegen das Klonen

Verlinsky Y., Strelchenko N., Kukharenko V., Rechitsky S, Verlinsky O, Galat V, Kuliev A. Human embryonic stem cell lines with genetic disorders. Reprod Biomed 2005, 10

Vogel G. Stem cells. Oocytes spontaneously generated. Science. 2003 May 2; 300(5620):721. No abstract available. PMID: 12730567

Vogel G. and Holden C. Developmental biology. Field leaps forward with new stem cell advances. Science. 2007 Nov 23; 318(5854):1224-5. No abstract available. PMID: 18033853

Wade N. Pioneer of in Vitro Fertilization wins Nobel Prize. The New York Times, October 4, 2010

Waggoner T. Renowned U.K. Stem Cell Scientist Moves to France: Says U.K. Too Focused on Embryonic Research. LiveSiteNews.com Friday October 24 2008 lifesitenews.com/ldn/2008/oct/08102405.html

Wahlberg D. Egg donors: Make a difference, and a little cash. Wisconsin State Journal Fri., JUN 27, 2008 b

Wahlberg D. Fertility Day 2: Treatments, trials and triumphs: In vitro has come a long way. Wisconsin State Journal JUN 30, 2008 - 11:21 AM 2008, a

Wahlberg D. $8.9 million for UW's stem cell research. Three Universities were awarded federal stem-cell grants. Wisconsin State Journal 5. August 2008 Page D-D2, c

Waller S.A., Paul K., Peterson S.E. Hitti J.E. Agricultural-related chemical exposures, season of conception, and risk of gastroschisis in Washington State. Am J Obstet Gynecol, 2010 Mar; 2002 (3): 241 e 1-6

Wandtner R. Wer schützt Affen vor der Forschung? Zum Streit über die Bremer Tierversuche 9.December 2008, Frankfurter Allgemeine Zeitung

Ward S. Gulf "dead zone" predicted to set a record. The Advocate News, Baton Ruge, LA. Jul 16, 2008

WARF (Wisconsin Alumni Research Foundation): European Patent Office Decision on WARF Appeal for Stem Cell Patent. (Nov 27, 2008) Statement from Wisconsin Alumni Research Foundation www.warf.org/news/news.jsp?news_id=236

Warkulwiz V.P. The Doctrine of Genesis 1-1, iUniverse book 2007, ISBN: 978-595-45243-9, p. 405

Warkus M. Nicht Mensch, nicht Tier, nicht Sache. Zum moralischen Status fremdartiger Wesen. Verlag: GRIN, ISBN: 3638724093, 2005

Washburn J. The legal lock on stem cells. Two patents that cover key research areas are setting back science. New America Foundation, Los Angeles Times. April 11 2006

Watt H. Pre-Implantation Diagnosis, The Linacre Center for Healthcare Ethics, London NW8 9 SE, England, 1997

Weber M. Stanford Encyclopedia of Philosophy Max Weber. First published Aug 24. 2007, plato.stanford.edu/entries/weber/#EthConRes

Weiss R. "Scientists Announce Mad Cow Breakthrough", The Washington Post. Retrieved on 2007-01-01.

Weiss R. Conservatives draft a "bioethics agenda" for president. The Washington Post; 2005 March 8, Sect A: 6

Wells H.G. * 21. September 1866 in Bromley, Kent; † 13. August 1946 in London) Die Insel des Dr. Moreau, 1896

Wertz D. Germ-line Therapy Enters the Foreseeable Future. The Gene Letter. 1997 3(1) Obtained from: http://www.geneletter.org/0898/germ-line.htm

Westhead R. Troubling questions surround surrogate-born children in India. The Star (Toronto Edition) April 26, 2010

White H. Obama accused of false distinction between "therapeutic" and "reproductive" cloning. His comments do not rule out "clone and kill" legislation. LifeSiteNews.com, Life, Family and Culture News, Thursday March 12, 2009

Whittemore A.S., Harris R., Itnyre J. Characteristics relating to ovarian cancer risk: collaborative analysis of 12 US case-control studies. IV. The pathogenesis of epithelial ovarian cancer. Collaborative Ovarian Cancer Group. Am J Epidemiol. 1992 Nov 15; 136 (10): 1212-20. PMID: 1476143

Wiedemann P.M., Simon J., Schiktanz S. und Tannert C. EMBO REPORTS: Science and Society. The future of stem cell research in Germany. A Delphy study 2004

Wilesmith J.W., Wells G.A., Cranwell M.P., Ryan J.B. Bovine spongiform Encephalopathy: epidemiological studies. Vet Rec, 1988. 123: p. 638-644.

Wilesmith J.W.,Ryan J.B., Atkinson M.J. Bovine spongiform encephalopathy: epidemiological studies on the origin. Vet Rec, 1991. 128: p. 199-203.

Wilesmith J.W., Ryan, J.B., Hueston W.D. Bovine spongiform encephalopathy: case-control studies of calf feeding practices and meat and bone meal inclusion in proprietary concentrates. Res Vet Sci., 1992. 52: p. 325-331.

Willhardt R. Neuer Weg zu einer Alzheimer-Therapie, Eine neue Substanz kann die Alzheimersche Krankheit zumindest im Tierversuch wirksam bekämpfen. Nov.18 2008, Informationsdienst Wissenschaft, idw-online.de/pages/de/news289198

Williams et al. US Patent 6200606 - Isolation of precursor cells from hematopoietic and non-hematopoietic tissues and their use in vivo bone and cartilage regeneration, March 13, 2001 Patent Storm, patentstorm.us/patents/6200606.html

Willke J.C. Why can't we love them Both. Chapter 31, Unwanted, Heritage House '76 http://www.abortionfacts.com oth_31.asp \ online book 2006

Wilmut J. Centre for Regenerative Medicine (Zentrum für regenerative Medizin) der Universität von Edinburgh, Rosalin Institute Scotland,Telegraph 16.11.2007

Winter R. India's Scared Cow, Archaeology Online 2005-2008

Wiwo Klontier-Produkte spalten Amerika, US-Gesundheitsbehörde, 07.01.2008

Wloka G. Warum ich keine Anti-Baby-Pille verschreibe, In: Empfängnisverhütung. Fakten, Hintergründe, Zusammenhänge, Hrsg. v. SÜSZMUTH, Roland, Holzgerlingen 2000, 1131-1141

Yoon Y.S., Wecker A., Heyd L., Park J.S., Tkebuchava T., Kusano K., Hanley A., Scadova H., Qin G., Cha D.H., Johnson K.L., Aikawa R., Asahara T., Losordo D.W. Clonally expanded novel multipotent stem cells from human bone marrow regenerate myocardium after myocardial infarction. J Clin Invest. 2005 Feb;115(2): 326-38.PMID: 15690083

Zabarenko D. Gulf of Mexico "dead zone" overlaps BP spill zone. Washington Post, Monday, August 2, 2010

Zenit The World seen from Rome. Pope Benedict IX: Science isn't to be worshipped or feared. Calls for Greater Synthesis with Philosophy. Vatican City, Oct. 28. 2010 ZE10102809

Zhang S.Q., Li X.J., Johnson A. and Pankratz M. T. Human embryonic stem cells for brain repair? Philosophical Transactions of the Royal Society B Biol. Science 2008 January 12; 363 (1489): 87-99

Zhang Q., Itagaki K., Hauser C.J. Mitochondrial DNA is released by shock and activates neutrophils via p38 map kinase. Shock. 2010 Jul;34 (1):55-9.

Zhang Q., Raoof M., Chen Y., Sursal T., Junger W., Karim B., Itagaki K., Hauser C.L. Circulating Mitochondrial DAMPs Cause Inflammatory Responses to Injury. Nature. 2010 March 4: 464 (7285): 104-107

Zinkant K. Koreanischer Zufallstreffer. Der Klonbetrüger Hwang Woo-Suk arbeitete so schlampig, dass ihm aus Versehen eine Jungfernzeugung gelang. Die Zeit 8.März 2007, Seite 32

Zepp-LaRouche, Schillers Idee des Erhabenen: Lehren für die Herrschenden heute. Rede bei der Sommerakademie, Aus der Neuen Solidarität Nr.44 2001 19 August 2001

Zhang S., Shpall E., Willerson J.T., Yeh E.T.H. Fusion of Human Hematopoietic Progenitor Cells and Murine Cardiomyocytes Is Mediated by α4β1 Integrin/Vascular Cell Adhesion Molecule-1 Interaction. Cellular Biology, Circulation Research 2007; 100:693

Zhou Q., Brown J., Kanarek A., Rajagopal J., Melton D. A., In vivo reprogramming of adult pancreatic exocrine cells to beta-cells. Nature 2008, Oct. 2; 455 (7213): 627-632

Book Cover:

Slaughterhouse: Vegetarismus.ch, Freeware pictures in high resolution. Cows: 5.jpg Half pigs' and cows' head http://www. vegetarismus.ch/bilder/5.jpg (top left). Painting's: Charles-Joseph Natoire (top right). Babel/ Pieter Bruegel (bottom left). Cornfield: E. Breburda (bottom right) Designed: E. Breburda

All pictures © in this book are taken or designed by E. Breburda

Summary

"PROMISES OF NEW BIOTECHNOLOGIES" features current hotly debated benefits and drawbacks of altering the genetic make-up in phyto-, veterinary- and biomedicine. It is a comprehensive introduction to the most important technology of genetic manipulation that bears worldwide social consequences. The author carefully researched independent scientific studies and revealed the shocking and alarming reality behind genetic manipulations and the patenting of life forms. She argues with environmental activists, religious organizations, lawyers, politicians, philosophers, public interest groups, professional associations and other scientists to explain utilized scientific methods of biogenetic engineering. She disclosed unintended side effects in the environment, animals or humans. By doing so, she points out the scientific dilemma, which makes her book unique. Genetically modified organisms contaminate the ecological system and influence animal and human health. Contraceptives, used by humans, cannot be filtered from water and sewage treatment plants. Feminizing chemicals are making their way into our drinking water and contaminate all water living creatures. Male fish start to produce immature eggs in their sex organs. Female fish develop liver and kidney malformations and slow down the reproductive cycle. The genetic innovation of crops may even cause the death of bees, the most important pollinators. Can progress become regress without the proper knowledge of nature's actions? Do we recklessly disregard "environmental health" for the purpose of scientific "dreams," like immortality, eternal life, a cure for every disease and the elimination of all suffering? We forget that a genetic predisposition may only contribute to the likelihood of developing a disorder. Environmental triggers, lifestyle and epigenes are what actually make the individual sick. Does science back up the wrong horse by prioritizing human embryonic stem cell research? What are the alternatives and why are they not employed? Why do scientists argue they "depend" on human embryonic stem cell research as the "Golden Standard"? This engaging and easy to read book is meant for the layperson to gain understanding of the complexity of the "biotechnological revolution". It provides the reader with tools and compelling arguments to be able to participate with ease in the most discussed topics in our times.

About the Author

Edith E. M. Breburda, D.V.M., Ph.D. was born 1964 in Munich, Germany. From 1983-1995 she studied medicine, veterinary medicine, and did some undergraduate studies in agricultural-sciences and psychology at the Ludwig Maximilians University of Munich, the Free University of Berlin and Justus Liebig University of Giessen.

She received her D.V.M. from Justus Liebig University of Giessen/Hesse/Germany in 1995. 1996, she successfully completed a top graded Ph.D. at Philipps University of Marburg, Medical School with a therapeutic changing study in limb lengthening and cartilage healing.

Dr. Breburda is internationally recognized as outstanding scientist with exceptional research contributions in the area of biomedical research. Her training and research skills are unique. Her superb contributions are vital and have benchmarking impact on the US biomedical research efforts. She has over 15 years' experience of academic teaching and basic science research at internationally renowned universities in Germany and at the University of Wisconsin/Madison. She has been recognized in peer-reviews journals for her outstanding skills and unique contributions to biomedical research. She has published in high-ranking journals, two children books and one book about biotechnology in German.

www.ingramcontent.com/pod-product-compliance
Lightning Source LLC
Chambersburg PA
CBHW060331200326
41519CB00011BA/1901